通派淮揚菜

法式擺盤的藝術結合

Nantong-style Huaiyang Cuisine

Meets the art of French plating

陳萍 著

萬里機構

父母感言

説起女兒，我們的感覺就是心疼。照顧老的，照顧小的，她也樂此不疲。即使病了也會樂呵呵的説：「很快就會好了！」女兒總是怕我們二老操心，而事實上，女兒像極父親的內在堅強、剛毅和果斷，像極母親的操心，還特別像極我們的善良、樸實和勤勞。

女兒總是後知後覺。在小學、初中時，她成績很差，也不知道是甚麼原因，初二時突然變成好學生模樣，幾乎名列前茅。女兒到了 30 歲不結婚、不談戀愛，一心一意埋首在工作上；但到了某天，説結婚就結婚了。女兒總是想到甚麼就做甚麼，好像思考不多，女兒説等到思考完了，黃花菜都涼了，更何況哪有可能思考就有結果？慢慢的，我們二老也從操心到習慣與支持。

女兒説要出書，沒過兩天就看到亞洲著名餐飲作家方曉嵐老師、著名法廚 Ricky 老師、香港萬里機構 Karen 老師，還有餐飲界著名攝影師梁細權老師，在南通進行新書的拍攝工作。

女兒計劃總共出版九本著作，第一本是我們的家鄉菜——通派淮揚菜。我們對女兒説讓我們來寫個序，來表達對淮揚菜的尊敬，對通派淮揚菜的地方特色表示敬仰。女兒愛國、愛南通，一直以自己是中國人、南通人而驕傲，為女兒的用心點讚！我們兩個老的除了給予女兒無限的愛，總希望女兒先照顧好自己，再去工作、照顧他人。

感謝所有支持女兒的朋友們，有時間來南通作客，你們會愛上南通的。南通位於江海交匯之處，是教育之鄉，能人輩出，更是紡織之鄉、建築之鄉，土地肥沃，資源十分豐富。

陳萍父親、母親

二零二五年四月二十七日於南通

推薦序

認識萍姐是在 The Food Story 的廚房。七年前，萍姐前來打工，是兼職形式。她禮貌、熱情，稱呼我為「師傅」，一直到現在。廚房的工作她自然搶着去做，打工期間，她總是第一個到達廚房，直到最後打烊才跟工友一起離開，她自然地練就了一身本領，對自己及對食材要求嚴格。到後來才知道，萍姐是個有錢人，但她十分隨和，能和同事打成一片，大家都習慣稱呼她為「萍姐」，還有年輕同事稱呼她為「乾媽」。

萍姐邀請我到她的工作室，每次都有嶄新的發現。萍姐研究川菜、川味火鍋底料，研究江浙菜、廣東菜多年，非常專業，她做事非常有程序，擁有很清晰的菜譜搭配、裝盤技巧和製作流程，忙中有序，繁而不亂。

2020 年，萍姐修讀「Institut Disciples Escoffier 法國烹飪藝術文憑課程」及「Institut Disciples Escoffier 法國糕餅烹飪藝術文憑課程」，自然地將傳統正宗的中餐融合新學派中，成就了她個人的獨特菜系，在傳統正宗菜式中顯露新的元素。

萍姐是我最尊敬的朋友之一，為人處世豁達大方，工作極其認真，竭盡全力將每道菜式完美地演繹出來，我相信萍姐《通派淮揚菜》的菜式及製作方法，將會給讀者耳目一新之感！

張錦祥 Ricky Cheung

張錦祥 Ricky Cheung

- 「合萍共廚・共用藝術餐廳」HPGC 地中海菜系及萍廚辣辣菜系技術總監及導師
- 法國 Le Mieux Bistro 私房菜餐廳及西餐廳 Bistro Bon 創始人
- 曾任香港高級餐飲集團 The Food Story 集團行政總廚、萬麗海景酒店行政總廚
- 電視及電台飲食節目主持人、食譜專欄作家

序言

在撰寫這本書時，我的心中充滿了無限的感慨與感激。

幾乎所有人都曾問我一個問題：「你投資在餐飲行業這麼多錢，甚麼時候才能回本？」我總是樂呵呵地説：「這一切，皆是老天的安排；而我從未違逆這份命運的指引。」

我感恩我的父親、母親養育之恩，他們不僅賦予我生命，更以兩句説話為我的人生鋪設了堅實的基石。

父親教會我一句話：「事到臨頭，伸頭是一刀，縮頭是一刀，要勇敢的把頭伸出去。」這句話，蘊含了無盡的智慧與勇氣，它教會我在面對困境時，要有破釜沉舟的決心，無畏無懼地迎接挑戰，這讓我在創業的道路上，一次次跨越難關，不斷前行。

母親教會我一句話：「不要在草堆中捉蛇。」這句話以其獨特的哲理，成為了我人生中快樂的源泉。蛇在草堆中很安詳，它只是取暖，幹嘛要驚動它？我習慣管理自己，堅信只要自身擁有足夠的誠信、善良、勤奮和優秀，身邊的人與事自然向着美好的方向發展。

如今，父親和母親年邁，依舊酷愛以健康方式享受美食美酒。早些年，弟弟説：「我們在生我們、養我們的地方開家餐廳吧！」這是弟弟的老本行，他於九十年代中期創辦了連鎖酒店久久酒樓。我熱愛美食，喜歡烹飪，2018 年決定投入健康餐飲，傳承健康美食，過了不久創辦了「CP

萍廚」。這三十年來，我走南闖北，從中國香港及澳門等地區，到新加坡、馬來西亞，以及遙遠的歐洲、美國，我非常希望能將全球自然健康的飲食文化發揚光大。

這七年多以來，我系統地學習中外餐飲，畢業於法國知名廚師學院 Institut Disciples Escoffier，並獲得「Institut Disciples Escoffier 法國烹飪藝術文憑課程」及「Institut Disciples Escoffier 法國糕餅烹飪藝術文憑課程」雙學位證書，兼榮獲 Disciples Escoffier International 法國埃科菲廚皇協會紅帶授勳，擔任香港國際廚藝交流協會「中國事務幹事」。此外，我在 2024 年新加坡國際美食挑戰賽，獲得唯一團隊金獎與個人主菜類金牌。

除了在學院修讀文憑課程，我追隨 Ricky 張錦祥等大師級老師，系統地實操法國菜、港粵菜、辣辣菜、中西式糕點等菜系，在餐廳的工作經驗令我眼界大開。我終於了解當一道菜好吃又好看時，結果將是多麼的驚人，我也終於知道只要緊緊抓住基本訣竅，每個人都可以只花少許力氣，就能將餐點變成視覺盛宴。

我希望這本書，能將傳統八大菜之一的經典淮揚菜相關知識和技巧傳遞讀者，並傳授與之相關的法式淮揚菜擺盤技巧，讓讀者在自家廚房、商業廚房自由發揮。

這本書除了給讀者靈感和啟發想像外，更像各種設計和組成元素一樣，讓讀者可以靈活運用書內提供的材料，自行擷取、組合，在菜式裝盤中盡情發揮個人創意。

祝大家胃口大開，盡情嘗試各種烹飪技巧和擺盤方式！

陳萍

二零二五年四月二十七日於南通

陳萍

「合萍共廚・共用藝術餐廳」HPGC 創始人，自幼浸潤南通江海文化的開闊堅韌。早年赴港打拼，在建築與食品行業嶄露頭角後，毅然追尋味覺初心，遠赴法國頂尖廚師學院 Institut Disciples Escoffier 苦修六年，獲得法國烹飪藝術與法國糕餅烹飪藝術雙學位證書，更榮膺法國埃科菲廚皇協會紅帶授勳及多項國際金獎。

南通「教育之鄉」的基因，與陳萍致力於美食教育傳承的理念不謀而合。

- 物產為基：借力於長江三鮮、呂四海產等自然饋贈，踐行「零添加劑」健康承諾。
- 融合創新：將法式技藝注入通派淮揚菜，雕琢南通美食新名片。
- 薪火相傳：創辦 HPGC 廚藝學院，培育兼具國際視野的烹飪人才。

她創立的「合萍共廚・共用藝術餐廳」，打造四大板塊，將健康藝術生活哲學傳遞全球。

1. 共用廚藝學院：提供多樣化的廚藝教育，致力培養具國際視野的廚藝專才和廚藝愛好者。
2. 共用藝術餐廳：堅持零添加劑健康食材，是未來學院學生的實習場所。
3. 共用貿易：線上線下銷售健康食材與藝術產品。
4. 餐飲文旅：以江海文化為圓心，打造「從餐桌到自然」的特色體驗。

她帶領團隊在新加坡國際美食賽摘得個人主菜與團隊金牌，開設的大師課程讓「紮根南通，心懷世界」的飲食哲學香飄寰宇。

擺盤的技巧和樂趣

創意擺盤是運用不同的元素，巧妙地做出各種變化，將餐盤變成完美藝術作品的過程。

第一個步驟是功能表規劃；

第二個步驟是選擇合適的餐具和擺飾；

第三個步驟是擺盤。

在 Institut Disciples Escoffier 就讀期間，令我印象最深的是 Mise en place，法語意思是——各就各位，或一切就緒。我認為這是優秀餐飲界的神奇公式，也是我的經驗之談。

首先將食譜從頭到尾認真讀一遍，並排列出各個組成部分的最佳烹飪時間，然後為所有工作步驟做好準備——放好廚房用具；食材按照所需分量洗切妥當。擺盤前，可將所有同屬一個盤子的食材及組成部分放在小托盤；烹飪完畢必要時需要加熱；醬汁及泥狀食物放進尖嘴擠壓瓶；所

* 廚師用香草及食用花裝盤擺設，將餐膳化成一道道藝術作品。

有香草用打濕的廚房紙巾包裹起來。烹飪過程中，廚房可能會像戰火摧殘過，要隨時擦拭和收拾。如果希望食物保持原有的溫度，碗碟應提前預熱或冰鎮。這一切都是 Mise en place 的好處。

美食的享受是從腦袋開始，並非從嘴巴開始。看到精緻擺盤的餐盤食物，你會自然想像食物的美味，會不知不覺地流涎。不管是好好的犒勞自己，烹煮一餐浪漫的晚宴，還是在家人或親友面前展示你的烹飪技藝，餐桌和擺盤作為第一步的視覺效果，會直接影響食慾和心情。

優雅的餐具、經典的香草配飾只是第一步，若想一頓餐膳變成難忘的經歷，我們要做些計劃和準備。

正如書內菜式——冰糖蹄膀，普通的慢烤櫻桃番茄、醃漬洋蔥芯、豆苗、濃縮的蹄膀原汁，通過主料的疊加層次感，裝飾界限劃分，醬汁銜接，無論點、面、空白相結合。

掌握了基本原則，加上本書的引導，稍稍的練習，你可以做出擺盤的效果。本書我所選擇的食譜，從廚房新手的簡單食譜，到給老手的複雜食譜一應俱全，大家快樂的開始吧！

目錄

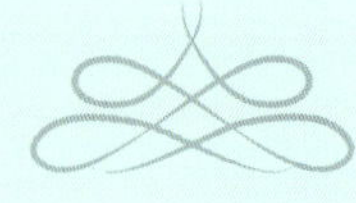

前菜 *Appetizers*

江鮮及河鮮 *Freshwater Fish and Shellfish*

富貴河中鮮 60
Lavish Braised Puffer Fish

紅燒春刀魚 64
Red Braised Chinese Tapertail Anchovy

刀魚魚丸 67
Chinese Tapertail Anchovy Fish Balls

文蛤燉蛋 68
Egg Custard with River Clams

白汁野生鮰魚 71
White Braised Wild-caught Long-snout Catfish

清蒸太湖白條 74
Steamed Sharpbelly from Lake Taihu

薑葱爆河蝦 78
Stir Fried River Shrimps with Ginger and Spring Onion

梁溪脆鱔 81
Liang Xi Crispy Eel

魚米之鄉 84
The Land Of Plenty (Sautéed Diced Mandarin Fish with Sweet Corn Kernels)

松鼠鱖魚 88
Squirrel-shaped Mandarin Fish

年糕燜毛蟹 92
Braised Hairy Crabs with Sticky Rice Cake

蒜子燜鰻魚 96
Braised Eel with Garlic Cloves

鯽魚羊方 99
Lamb Stuffed with Crucian Carp

荷包蒸桂花魚 102
Steamed Mandarin Fish in Chicken Stock Glaze

海鮮 *Seafood*

小米遼參 Spiny Sea Cucumber in Millet Porridge	108
米湯灼花膠 Fish Maw Blanched in Rice Soup	112
乾燒大明蝦 Dry-braised Jumbo Shrimps	115
酥香帶魚 Deep-fried Hairtail	118
香酥梅子魚 Deep-fried Big-head Croaker	122
金湯黃魚脯 Yellow Croaker Fillet in Golden Broth	126
蝦仁鍋巴 Shrimps on Scorched Rice	129
蘇式鯧魚 Jiangsu-style Smoked Pomfret	132
大烤墨魚 Braised Cuttlefish	136
絲瓜文蛤蟶乾湯 River Clam and Dried Razor Clam Soup with Angled Loofah	139

禽肉 *Meat & Poultry*

上海八寶雞 Shanghainese Eight-treasure Chicken	144
冰糖蹄膀 Braised Pork Knuckle in Rock Sugar Soy Sauce	148
紅燒海門羊肉 Red Braised Haimen Goat	151
醃篤鮮 Yanduxian (Cured Pork and Fresh Pork Soup with Bamboo Shoot and Tofu Skin Knots)	154

八寶雞翅 158
Eight-treasure Stuffed Chicken Wings

蟹粉獅子頭 162
"Lion Head" Meatballs with Crabmeat and Crab Roe

芋艿清湯鴨 165
Duck Soup with Yam

農家蒸虎皮肉 168
Tiger-striped Pork Belly

通城藕餅 172
Tongcheng Fried Lotus Root Stuffed with Pork

霸王別姬 175
Hegemon-King Bids His Lady Farewell (Chicken and Softshell Turtle Soup)

點心及豆品 *Dim Sum & Soybean Products*

蟹黃鮮肉蒸餃 180
Steamed Dumplings with Pork and Crab Roe

黃橋燒餅 184
Huangqiao Scallion Flatbread

草頭煎餅 187
Pan-fried Burclover Dumpling

翡翠燒賣 190
Jadeite Shaomai

薺菜餛飩 194
Shepherd's Purse Wonton

水果酒釀元宵 198
Mini Glutinous Rice Balls with Fresh Fruits in Jiuniang Sweet Soup

菊花豆腐 201
Chrysanthemum Tofu

大煮乾絲 204
Braised Shredded Dried Tofu

古今淮揚菜

淮揚，指的是淮海和揚州，淮揚菜是中國最歷史悠久的菜系之一，在中國八大菜系中，代表的就是江蘇菜。

據《清稗類鈔・飲食類・各省特色之餚饌》中所述，「餚饌之有特色者，為京師、山東、四川、廣東、福建、江寧、蘇州、鎮江、揚州、淮安」。在此十個地方菜中，江蘇省佔五席，可見古今淮揚菜之影響深遠。

江蘇地區，物產豐饒

江蘇省地踞長江、淮河南北，平原、丘陵、湖泊、河流一應俱全。江蘇省四季分明，雨量充沛，土地肥沃，物產豐富，得天獨厚。由於交通方便，經濟發達，人們生活條件富足，距今七千年前的浙江余姚河姆渡文化，說明當時長江下游地區已人工種植水稻。

「飯稻羹魚」的魚，是指品種繁多的水產，包括魚、蝦、蟹、蛤、螺、蚌、甲魚等。據記載，在夏商周時期，江蘇地區已經懂得養魚技術，而在江蘇睢寧、銅山的漢代石畫上，有記載當時百姓捕魚的場面。江蘇省水網交織，並擁有太湖和陽澄湖等大湖泊，生態條件得天獨厚，水產非常豐富，尤其江蘇清水大閘蟹更是名聞中外，而以蟹入饌更是江蘇名菜和名點（例如大閘蟹、蟹粉豆腐、蟹粉小籠包等）。

南北交流，技術文化相融

東漢後期，中原大地經歷黃巾起義、董卓之亂、魏蜀吳三國相互征戰，加上五胡亂華等戰爭，社會動盪不安，造成大量北方人舉家南遷，其中包括了不少名門望族。南北朝時代的政權，把他們安頓在「南徐州」和「南兖州」，設「僑郡、僑縣」，即今天常州、鎮江、揚州、淮安等地區。北方南遷的人口，為南方帶來了人力、資金和生產技術，更重要的是帶來了麥、菽、粟的種植和加工技術，並在江蘇地區推廣起來。魏晉時期，江蘇地區糧食加工技術已很進步，稻米加工成精細的白米，而且發明了絹羅製的麵粉篩子，正如晉代束晢《餅賦》中曰「重羅之麵，坐飛雪白」。精細麵粉的出現，大大促進了江蘇地區的點心類食物發展。

江蘇省在古代屬於淮揚地區，「淮揚」的名稱從商周時代已經出現，清康熙六年才正式被分為江蘇和浙江兩省及安徽的一部分。清代康熙、雍正、乾隆三朝盛世，長江下游地區是國家經濟中心，鹽茶漕運、絲綢紡織，為淮揚地區帶來前所未有的興旺，是古代淮揚菜發展的頂峰，飲食文化交流，加上南遷的豪門和名士帶來京都的烹調技術，中原飲食文化和吳越飲食文化交匯融合，使江南菜餚更加多樣化和精細。宋元時期，社會經濟快速發展，特別是宋朝時期開放夜市，酒樓食店如雨後春筍，飲食業一片繁華景象，百花齊放。

❊ 文人雅士，推崇江蘇飲食文化

到了清代，江蘇省由「魚米之鄉」進一步成為「天下糧倉」。康熙和乾隆曾多次南巡，對江蘇地區的飲食文化影響很大。公元 1780 年（乾隆四十五年）乾隆第五次南巡到揚州，百官為他設的「滿漢席」，即滿席和漢席合併，菜餚極為豐盛，而且寓意「滿漢官民，俱為一家」，為後世留為佳話，對江南飲食文化有很大的促進作用。

江蘇省人傑地靈，自古就是風流才子文人雅士的集中地。造就了江蘇人對飲食的講究程度，高於我國較遲開放的省份，據《史記》、《吳越春秋》記載，江蘇人在兩千多年前已懂得用蒸、炒等技法來烹調魚類，以保持其鮮味；而在千年以前的宋朝時期，金陵（南京）鹽水鴨和烤鴨，已經遠近馳名。中國古代的十大名廚中有五位來自江蘇和浙江，他們是太和公、劉娘子、蕭美人、王小余和家喻戶曉的董小宛，而蘇東坡、陸游、袁枚、曹寅、唐伯虎等，既是文人也是美食家，曹雪芹在《紅樓夢》提及的飲食，指的也是淮揚菜。尊敬的開國總理周恩來是淮安人，1949 年 10 月 1 日北京開國大典前夕，毛澤東等領導人宴請各界人士，在周恩來的提議下，宴會的菜式選的就是江蘇淮揚菜。

❊ 帝王下江南，提升揚州菜水平

淮揚菜的中心在揚州市，這個地區在春秋時代叫廣陵郡，到了漢朝曾多次改為廣陵國，後又改回廣陵郡。在晉朝時和京口（今鎮江）同是北府兵的兩個駐兵地點。隋朝曾名為江都，到了唐朝才改名為揚州。隋唐時代稱為淮揚的地區，以江都（揚州市）為經濟中心，這裏集中了揚州和淮安地區美食的精華。揚州人以其飲食文化為榮，至今仍稱揚州菜為淮揚菜。

一千四百年前，隋朝京杭大運河開通，隋煬帝三下揚州，宮廷的菜式進入淮揚地區，進一步提高了揚州菜的烹調水平。大運河從洛陽途經江都（今揚州）伸延到餘杭（今杭州）。元、明、清三代，繼續修建北到天津，西連長安。大運河也為各朝的皇帝提供了出巡的方便，隋煬帝三下揚州，康熙六次南巡，乾隆六下江南，留下了不少風流韻事，古時由南至北的漕運主要依靠大運河，揚州於是成為鹽商聚居的城市，《資治通鑑》說「揚州富庶甲天下，時人稱揚一益二」（益即四川），揚州人有錢，當然對享受更精益求精，所以才有揚州三把刀：菜刀、理髮刀、修腳刀。就菜刀而言，如果沒有揚州精湛的菜刀工藝，恐怕也不會有精致的揚州菜。

當年隋煬帝楊廣帶着妃嬪隨從，乘着龍舟沿大運河南下看瓊花（即繡球花），「所過州縣，五百里內皆令獻食」。楊廣對萬松山、金錢墩、象牙林、葵花崗等四大名景十分留戀。回到行宮後，楊廣吩咐御廚以上述四景為題，製作四道菜餚作為記念。御廚們費盡心思終於做成了松鼠桂魚、金錢蝦餅、象牙雞條和葵花斬肉（即獅子頭）四道菜。隋煬帝品嘗後甚悅，賜宴群臣，令淮揚菜風行朝野。

❊ 揚州菜手工精巧，名菜獨特

淮揚菜的廚藝被列為全國之首，今天的揚州是全國著名的烹飪教育中心。揚州菜的特色是手工細膩，菜式清爽雅淡，薄油輕獻，講究鮮嫩軟滑，做法以煨、燉、燴、炒為，其特色是擅長做湯菜。一塊白豆乾，刀工高明的揚州師傅能夠橫片出二十四層（也有說二十八層）厚薄平均的豆乾片，再切成比牙籤更細的乾絲，根根一樣粗細，煮不爛、理不亂，和火腿絲、雞絲同煮，可以把三絲均勻地混在一起，每一口都能夠吃到三絲。當年乾隆皇下江南，揚州官員奉上「九絲湯」，是以乾絲加上八種其他切成絲狀的食材所做成。清嘉慶年間揚州鹽商童岳薦選撰的《調鼎集》，記載了「九絲湯」所用的材料：火腿絲、筍絲、銀魚絲、木耳、口蘑、千張、腐乾、紫菜、蛋皮、青筍或加上海參、魚翅、蟶乾、燕窩等。後來，民間簡化了材料，成了今天著名的「大煮乾絲」或「雞火乾絲」。

淮揚菜選料新鮮丶不時不食、廚藝考究、刀工細膩見稱。揚州的著名菜式有「三頭」：揚州獅子頭、拆肉魚頭、冰糖煨豬頭；「三醉」：醉蟹、醉蝦、醉螺。其他還有著名的揚州炒飯、鹽水鵝丶文思豆腐、炒軟兜、紅燒江鰻、桂花糖藕等等，美味菜式多不勝數。

長江刀魚是江蘇名產，刀魚的形狀身薄如刀，腹部為銀白色，背部灰色。刀魚刺多細軟，肉質非常鮮嫩肥美，為江魚中之極品。野生長江刀魚棲息在近海處，每年三月中旬游入長江，在中下游的淡水河口產卵，蘇東坡說的「恣看收網出銀刀」，説的就是這一帶產的刀魚。

❉ 通派淮揚菜，別有風味

江蘇省南通市，經歷兩千年的歲月沉澱，是一座擁有文化傳統的城市，南通古稱通州，如皋是南通的一部分，這個地區在清朝中期之前屬淮揚府管轄，位置在上海的北面隔岸，是一片江海平原，平原上流淌的濠江是長江出海的支流之一。南宋末年，愛國詩人文天祥從南通渡江，寫下「春紅堆蟹子，晚白結鹽花」的詩句，表達對這片生機盎然的鄉村無限眷戀。

南通自古是南北交匯的魚米之鄉，四季物產豐富，河豚、帶魚、鯧魚、刀魚、狼山雞、蟶子，特別是號稱「天下第一鮮」的文蛤。南通的菜式屬淮揚菜，稱為「通派淮揚菜」，是淮揚菜中的一個流派，與揚州淮揚菜同出一轍，但風味不同，各有特色。

撰文：方曉嵐

香港飲食文化作家、食評家、報章專欄作家、香港中華廚藝學院大師級課程客座導師及論文評審。曾出版二十多本飲食文化食譜書，另擔任《香港志－飲食卷》總撰。

CHAPTER 1

Appetizers

前菜

In recent years, starting a warm meal with cold appetizers has become a norm. My childhood memories of cold appetizers like drunken chicken and drunken shrimps are especially vivid. I didn't find them especially tasty at first. But as I grew older, I started to grasp the life philosophies therein.

The brewing process of Chinese yellow wine began with rice grains which are then partly fermented into jiuniang. Dead yeast cells left after the fermentation are called wine lees. Jiuniang, wine lees and Chinese yellow wine are all excellent ingredients in Chinese cooking. Such transformations not only show the passage of time, but also give birth to a liquor known as Hualu Shao, famous in the Jiangzhe region. The liquor is used to make drool-worthy drunken seafood and river fish. Among them, drunken crab and drunken sharpbelly stand out with their flavoursome flesh and unique winey aromas. Drunken clams, declared the world's most umami-laden food, is considered the crème de la crème among all cold appetizers.

在當今社會，「冷菜溫食」已成為一種風尚，而我對兒時記憶中的醉雞、醉蝦尤為深刻。儘管初嘗時或許未覺得特別美味，但隨着年歲增長，這些菜餚蘊含的人生哲理卻逐漸顯現。

從酒釀到酒糟，再到黃酒，這一系列的演變不僅展現了時間的沉澱，更催生了江浙地區著名的花露燒，一系列的醉海鮮、醉河鮮令人垂涎。其中，醉螃蟹、醉白條以其鮮美的肉質和獨特的酒香脫穎而出，而被譽為「天下第一鮮」的醉文蛤，更是前菜中的精品。

鎮江水晶餚肉

Zhenjiang-style Pork Trotter Crystal Aspic

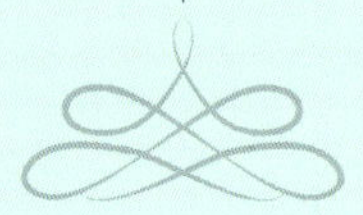

又稱為「水晶餚蹄」，
是江蘇傳統名餚，
迄今已有三百餘年歷史，
色澤鮮艷，晶瑩澄澈，
肥瘦相間，迴味悠長。

This Jiangsu classic has over 300 years of history.
It's popular not only because of
its blush colour and gem-like appearance,
but also for its perfect ratio of lean and fatty meat,
as well as its lingering flavours.

材料

豬蹄膀 _2 個

配料

花椒 _5 克
八角 _5 克
薑 _30 克
葱 _35 克
糖 _5 克
鹽 _200 克
花雕酒 _20 克

醋珠材料

大紅浙醋 _125 克
魚膠片 _2 片
橄欖油 _250 克

做法

1_ 花椒、八角及鹽放入鍋炒至香味散發，成為花椒鹽。

2_ 豬蹄膀清洗乾淨，去除骨，皮朝下，肉厚位劃上十字花刀；倒入糖及花椒鹽抹勻，醃 24 小時。

3_ 將醃好的豬蹄膀飛水，沖洗乾淨；放入不銹鋼盆，加入水蓋過肉面，加入薑、葱、花雕酒，蒸 90 分鐘，取出，皮朝下倒入原湯，壓上重物，置於冰箱冷藏一夜。

4_ 魚膠片放入冰水泡軟，擠乾水分，放進大紅浙醋，加熱溶化，待涼，用擠花袋擠入座於冰塊的橄欖油中成醋珠。

5_ 豬蹄膀取出，切成方塊，整齊地擺放碟內，撒上薑絲、醋珠及葱花即成。

Ingredients

2 pork trotters

Condiments

5 g Sichaun peppercorns
5 g star-anise
30 g ginger
35 g spring onion
5 g sugar
200 g salt
20 g Huadiao wine

Vinegar spheres

125 g red vinegar
2 gelatine leaves
250 g olive oil

Method

1_ Fry Sichuan peppercorns, star-anise and salt in a dry wok until fragrant. This is Sichuan pepper salt.

2_ Rinse the pork trotters. Debone them. Put them on a counter with the skin side down. Make light crisscross cut on the meat side. Sprinkle with sugar and Sichuan pepper salt from step 1. Rub evenly and leave them for 24 hours.

3_ Blanch the pork trotters in boiling water. Rinse well. Transfer into a stainless steel tray. Add enough water to cover. Put in ginger, spring onion and Huadiao wine. Steam for 90 minutes. Remove from heat and keep the pork trotters in the broth with the skin side down. Put heavy weight over the pork trotters to stop them from floating. Refrigerate overnight.

4_ To make vinegar spheres, soak the gelatine leaves in the iced water until soft. Squeeze the water and add to the red vinegar. Heat over low heat until gelatine dissolves. Leave it to cool. Transfer into a piping bag. Pipe the mixture one drop at a time into cold olive oil standing over the ice cubes. Strain the vinegar spheres with a fine-mesh strainer.

5_ Turn the pork trotter aspic out of the stainless steel tray. Cut into cubes. Arrange neatly on a serving plate. Sprinkle with finely shredded ginger, vinegar spheres, and finely chopped spring onion. Serve.

四喜烤麩

Four-happiness Braised Wheat Gluten

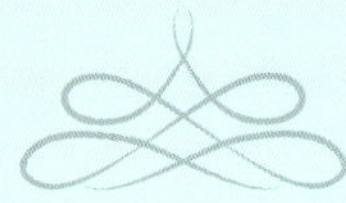

以金黃蓬鬆的烤麩為基底，
汁稠味厚而不膩，
口感多樣，象徵圓滿富足，
體現江南飲食濃油赤醬的一面。

Wheat gluten are fried till fluffy and golden.
The sauce is rich, but not greasy.
The varied textures represent affluence and completeness, embodying the quintessence of Jiangnan culinary culture – the use of ample oil with a mildly sweet sauce coloured reddish brown by soy sauce.

材料

上海乾烤麩 _110 克
毛豆仁 _25 克
乾小冬菇 _12 朵
乾雲耳 _10 克
乾金針菇 _20 克

配料

薑汁 _4 茶匙
薑絲 _20 克

調味料

蠔油 _2 湯匙
萬字醬油 _1 湯匙
草菇老抽 _1 湯匙
紹興酒 _2 湯匙
糖 _4 茶匙
麻油 _2 茶匙
水 _ 15 克

做法

1_ 上海烤麩切塊（如選用香港烤麩，撕成小塊）。

2_ 鍋內注入水 1 公升，撒入鹽半茶匙、薑汁 2 茶匙煮沸，放入烤麩飛水 1 分鐘，盛起，沖水後用手擠出水分，重複此步驟至無黏液，晾曬或低溫烤烘。

3_ 燒熱一鍋水，放入毛豆仁煮沸 5-6 分鐘，瀝乾水分。

4_ 雲耳浸泡滾水發好，放入沸水中煮 2 分鐘，瀝乾水分；冬菇浸軟，去蒂；金針菇浸 10 分鐘，用水沖涼，飛水備用。

5_ 熱鍋冷油，放入上海烤麩煸炒，加入紹興酒拌炒（如用香港烤麩，燒熱油 500 毫升，待六成熱，炸成金黃色，盛起瀝油，用鍋鏟壓出油分）。

6_ 燒熱油 6 湯匙，大火起鍋，放入薑絲、冬菇、毛豆仁、金針菇及雲耳爆香，加入紹興酒、蠔油、萬字醬油及草菇老抽，灑入餘下薑汁、鹽及糖炒勻，拌入烤麩煮至醬汁濃稠，上碟前加入麻油，擺盤即成。

Ingredients

110 g Shanghainese dried deep-fried wheat gluten
25 g green soybean kernels
12 small dried shiitake mushrooms
10 g dried cloud ear fungus
20 g dried day lily flowers

Aromatics

4 tsp ginger juice
20 g finely shredded ginger

Seasoning

2 tbsp oyster sauce
1 tbsp Kikkoman soy sauce
1 tbsp mushroom-flavoured dark soy sauce
2 tbsp Shaoxing wine
4 tsp sugar
2 tsp sesame oil
15 g water

Method

1_ Cut the Shanghainese deep-fried wheat gluten into chunks. (Tear it into pieces if you're using Hong Kong-style wheat gluten.)

2_ Pour 1 litre of water into a pot. Add 1/2 tsp of salt, and 2 tsp of ginger juice. Bring to the boil. Put in the wheat gluten chunks and blanch them for 1 minute. Drain and rinse under a running tap till cold. Squeeze them dry with your hands. Repeat this step until there is no longer slimy liquid coming out of the wheat gluten. Spread the gluten chunks out on a tray to air-dry. Or bake them in an oven at the lowest temperature till dry.

3_ Boil a pot of water and blanch the green soybean kernels for 5 to 6 minutes. Drain.

4_ Soak cloud ear fungus in water until soft. Blanch in boiling water for 2 minutes. Drain and set aside. Soak shiitake mushrooms in water until soft. Cut off the stems. Set aside. Soak day lily flowers in water for 10 minutes. Rinse well. Blanch in boiling water and drain.

5_ Heat a wok until hot. Add oil. Fry the wheat gluten chunks and toss well. Drizzle with Shaoxing wine and toss again. (If you're using Hong Kong-style wheat gluten, heat 500 ml of oil in a wok until 150 to 180°C. Deep-fry the wheat gluten until golden. Drain and squeeze out any excess oil with a spatula.)

6_ Heat 6 tbsp of oil in a wok. Stir-fry shredded ginger, shiitake mushrooms, green soybean kernels, day lily flowers and cloud ear fungus until fragrant. Drizzle with Shaoxing wine, oyster sauce, Kikkoman soy sauce, and mushroom-flavoured dark soy sauce. Add the remaining ginger juice, salt and sugar. Toss to mix well. Put in the wheat gluten chunks and cook until the sauce reduces. Drizzle with sesame oil before plating. Serve.

通城醉雞

Tongcheng-style Drunken Chicken

作為淮揚菜的經典美食，
醉雞的亮點在於其特製酒料，
通城醉雞選取的是花雕酒，
大家製作時可以選擇其他黃酒代替。

A representative of Huaiyang classics,
drunken chicken seduces with its unique wine
marinade infused with herbal aromas.
For this recipe, we use Huadiao in the marinade.
But you may replace it with your favourite
cooking wine instead.

材料

母雞 _1 隻（約 1.2 公斤）

配料

薑 _15 克
葱 _20 克
料酒 _50 克
鹽 _15 克

通城醉料

杞子 _50 克
參鬚 _80 克
甘草片 _15 克
川貝片 _15 克
紅棗 _150 克
水 _3 公斤
生抽 _120 克
鹽 _25 克
糖 _40 克
花雕酒 _80 克

做法

1_ 將通城醉料的香料放入水內煮開，關火，加入鹽、糖及生抽，冷卻後，加入花雕酒，放入冰箱冷藏。

2_ 雞去掉內臟，洗淨，擦乾水分。

3_ 雞隻表皮及內腔用配料抹勻，醃 1 小時。

4_ 雞放入蒸鍋蒸 30 分鐘，取出待涼，放入醉料內放置 12 小時。

5_ 雞切件，上碟，淋上少許醉料，放上參鬚、紅棗、杞子及川貝片點綴即可。

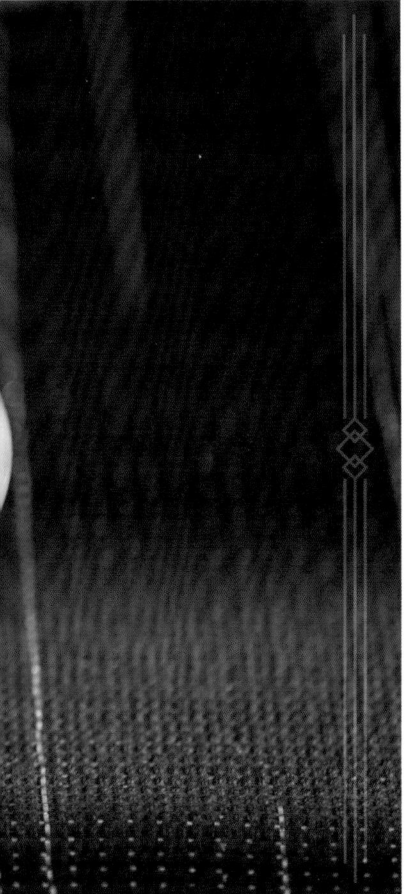

Ingredients

1 hen (about 1.2 kg)

Condiments

15 g ginger
20 g spring onion
50 g cooking wine
15 g salt

Tongcheng-style wine marinade

50 g dried goji berries
80 g ginseng fibrous roots
15 g sliced liquorice
15 g sliced Chuan Bei
150 g red dates
3 kg water
120 g light soy sauce
25 g salt
40 g sugar
80 g Huadiao wine

Method

1_ To make the wine marinade, put all spices and herbs into water. Bring to the boil and turn off the heat. Add salt, sugar and light soy sauce. Leave it to cool. Add Huadiao wine and keep it refrigerated.

2_ Remove the innards of the chicken. Rinse and wipe dry.

3_ Rub condiments evenly on both the inside and outside of the chicken. Leave it for 1 hour.

4_ Put the chicken on a steaming dish. Steam for 30 minutes. Leave it to cool at room temperature. Transfer into the wine marinade from step 1. Soak it for 12 hours in the refrigerator.

5_ Slice the chicken and arrange on a serving plate. Drizzle with a few tablespoons of the wine marinade. Garnish with ginseng fibrous roots, red dates, goji berries and Chuan Bei. Serve.

老滷三拼盤

Marinated Pork Platter

滷口條・滷豬耳・滷豬心

pork tongue・pork ear・pork heart

滷製技法始於先秦，
製作簡便，鮮嫩美味，
深受沿海一帶地區人們鍾愛，
為淮揚傳統名吃之一。

The use of marinade to infuse food with flavours began in early Qin Dynasty (around 220 BCE). It adds depth to food in a few simple steps, and it is hugely popular among people living along the coast. It is also one of the most famous cooking methods in the Huaiyang culinary tradition.

材料
口條（豬脷）_1 根（約 250 克）
豬耳 _1 個（約 300 克）
豬心 _1 個（約 250 克）

鹵水香料
香葉 _15 片
白芷 _15 克
草果 _15 克
砂仁 _5 克
肉蔻 _10 克
小茴香 _20 克
八角 _20 克
甘草 _5 克
花椒 _20 克

配料〈飛水用〉
薑片 _30 克
葱 _30 克
黃酒 _100 克

配料
豬肘子 _1 個
豬腳 _2 個
老母雞 _1 隻
雞腳 _1.2 公斤
生薑 _150 克
水 _18 公斤
生抽 _60 克
冰糖 _80 克
鹽 _400 克
白酒 _100 克

做法

1_ 鹵水香料用溫水泡 20 分鐘，洗淨瀝乾，包好放入香料袋。

2_ 豬肘子、豬腳、老母雞及雞腳洗淨血水；鍋內放入冷水，下薑、葱及黃酒，放入以上材料，飛水，洗淨瀝乾。

3_ 豬肘子、豬腳、老母雞、雞腳、鹵水香料及餘下配料（白酒除外）放入鍋內，大火煮開，加入白酒 50 克，轉小火煮 5 小時，取出所有材料，保留香料包，備用。

4_ 口條、豬耳及豬心飛水，洗淨瀝乾備用。

5_ 煮滾鹵水，放入口條、豬耳及豬心煮開，加入白酒 50 克，以中火煮 40 分鐘，再轉小火煮 20 分鐘，關火加蓋待 30 分鐘。

6_ 鹵肉取出，淋上已爆香葱油，待至常溫，修切裝盤點綴即成。

Ingredients
1 pork tongue (250 g)
1 pork ear (300 g)
1 pork heart (250 g)

Spices for marinade
15 bay leaves
15 g Bai Zhi
15 g Tsaoko fruit
5 g Sha Ren
10 g nutmeg
20 g fennel
20 g star anise
5 g liquorice
20 g Sichuan peppercorns

Blanching stock
30 g sliced ginger
30 g spring onion
100 g Chinese yellow wine

Marinade base
1 pork fore leg
2 pork hind legs
1 old hen
1.2 kg chicken feet
150 g ginger
18 kg water
60 g light soy sauce
80 g rock sugar
400 g salt
100 g Chinese rice wine

Method

1_ Soak all spices for marinade in warm water for 20 minutes. Rinse and drain well. Put them into a muslin bag and close it.

2_ To make the marinade base, rinse pork fore leg, pork hind legs, old hen and chicken feet. Put them into a pot and fill it with cold water. Add blanching stock ingredients. Bring to the boil and cook until foamy scum floats on top. Drain and rinse well under a running tap.

3_ Put pork fore leg, pork hind legs, old hen and chicken feet into a pot. Add the spices for marinade and all remaining marinade base ingredients, except Chinese rice wine. Bring to the boil over high heat. Add 50 g of Chinese rice wine. Turn to low heat and simmer for 5 hours. This is the marinade base. Remove all ingredients but keep the spice bag in.

4_ In another pot, blanch pork tongue, pork ear and pork heart in boiling water. Drain and rinse well.

5_ Bring the marinade base from step 3 to the boil. Put in the blanched pork tongue, ear and heart. Bring to the boil again. Add 50 g Chinese rice wine. Cook over medium heat for 40 minutes. Turn to low heat and simmer for 20 minutes. Turn off the heat and leave it for 30 minutes with the lid on.

6_ Remove pork tongue, ear and heart from the marinade. Drizzle with scallion oil (sauté scallion in the oil before). Leave them to cool to room temperature. Slice and arrange on a serving plate. Garnish and serve.

翡翠海蜇絲

Emerald Jellyfish Salad

以晶瑩剔透的優質海蜇為主角，
切細如絲，脆嫩彈牙，
是宴席冷盤的亮麗之品，
亦為家常餐桌的輕盈之選。

Translucent, high-quality jellyfish is shredded finely
to give this dish a bouncy texture.
This is a highlight among cold appetizers served in banquets.
It also makes a light,
healthy starter in any home-style dinner.

材料

海蜇皮 _200 克
萵筍 _100 克
紅蘿蔔 _50 克

調味料

鹽 _2 克
糖 _2 克
鮮醬油 _5 克
胡椒粉 _0.5 克
麻油 _5 克
熱油 _10 克

做法

1_ 海蜇皮切成細絲，用水沖約 30 分鐘，去掉鹽味和澀味，瀝乾水分，擠乾備用。

2_ 萵筍切成細絲，加入鹽 1 克醃 10 分鐘，擠乾水分，備用。

3_ 紅蘿蔔切成絲，放入熱水略灼，盛起，立即放入冰水，冷卻後，瀝乾水分。

4_ 海蜇、紅蘿蔔絲及萵筍絲放入大碗內拌勻，加入鹽、糖、鮮醬油及胡椒粉拌勻，最後拌入熱油及麻油，擺盤上碟即成。

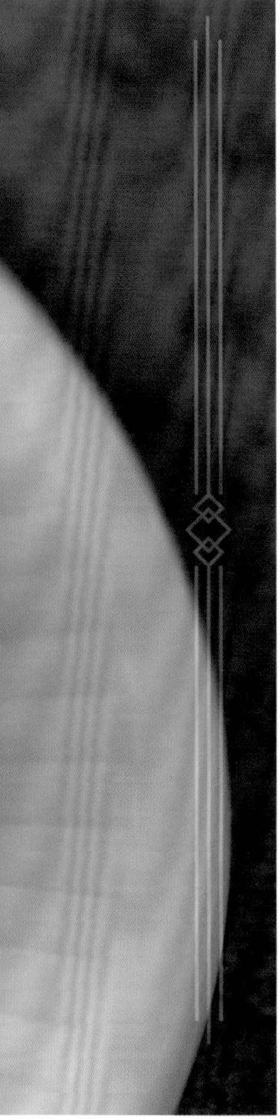

Ingredients
200 g jellyfish skin
100 g celtuce
50 g carrot

Seasoning
2 g salt
2 g sugar
5 g Maggi's seasoning
0.5 g ground white pepper
5 g sesame oil
10 g hot oil

Method

1_ Shred the jellyfish finely. Rinse in running water for 30 minutes to remove the salty and bitter taste. Drain well. Squeeze dry.

2_ Finely shred the celtuce. Add 1 g of salt and mix well. Leave it for 10 minutes. Squeeze dry and set aside.

3_ Finely shred the carrot. Blanch in boiling water briefly. Drain and soak in ice water to stop the cooking process. Drain.

4_ Toss jellyfish, celtuce and carrot in a mixing bowl. Season with salt, sugar, Maggi's seasoning and ground white pepper. Mix well. Stir in hot oil and sesame oil. Arrange on a serving plate. Serve.

糖醋梅花肉

Sweet and Sour Pork

糖醋梅花肉是清朝初年一道宮廷菜，
後來流傳民間，
成為淮揚菜系中重要的菜餚，
更是中國經典傳統名菜之一。

This is an imperial dish that
can be traced back to the early Qing Dynasty.
Later on, it became popular among peasants and
one of the most important Huaiyang recipes.
It's also one of the best known Chinese classics.

材料

梅花肉 _250 克
年糕 _100 克

配料

鹽 _2 克
粟粉 _50 克
薑片 _20 克
葱 _30 克
油 _100 克

調味料

上海紅醬油 _25 克
冰糖 _60 克
料酒 _30 克
清雞湯 _150 克
老陳醋 _50 克

做法

1_ 梅花肉切成 4 厘米正方塊，沖水 30 分鐘，瀝乾水分，灑入鹽和粟粉醃 30 分鐘。

2_ 年糕修切成 2.5 厘米正方塊，撲上粟粉，備用。

3_ 煎鍋內放入油 20 克，燒熱至 120℃，放入年糕塊煎至金黃色，取出備用。

4_ 燒熱鍋，放入油 80 克，加入薑及葱煸炒，燒熱至 120℃，放入梅花肉煎至表面金黃色，依次加入紅醬油、冰糖、料酒及清雞湯，煮滾後以小火燜 30 分鐘。

5_ 加入老陳醋 25 克收汁至半濃稠，放入年糕拌勻，待汁液至濃稠，倒入剩餘的老陳醋，翻炒均勻，起鍋裝盤即可。

Ingredients
250 g pork shoulder butt
100 g sticky rice cake

Condiments
2 g salt
50 g corn starch
20 g sliced ginger
30 g spring onion
100 g cooking oil

Seasoning
25 g Shanghai red soy sauce
60 g rock sugar
30 g cooking wine
150 g chicken stock
50 g aged vinegar

Method

1_ Cut the pork into 4-cm cubes. Rinse under a running tap for 30 minutes. Drain. Add salt and corn starch to the pork. Mix well and leave it for 30 minutes.

2_ Cut the sticky rice cake into 2.5-cm cubes. Coat them in corn starch. Set aside.

3_ Pour 20 g of cooking oil in a frying pan. Heat the oil to 120°C. Fry the sticky rice cake until golden. Drain and set aside.

4_ Heat a wok. Add 80 g of oil. Stir-fry ginger and spring onion until fragrant. Turn the heat up to 120°C. Shallow-fry the pork cubes until golden. Add red soy sauce, rock sugar, cooking wine and chicken stock in the order specified. Bring to the boil and turn to low heat. Simmer for 30 minutes.

5_ Add 25 g of aged vinegar and cook until the sauce reduces by half. Put in the fried sticky rice cake and toss until the sauce thickens further. Add the remaining aged vinegar. Toss well. Save on a serving plate. Serve.

金陵鹽水鴨

Jinling Brine-poached Duck

金陵鹽水鴨是六朝古都南京的標誌性美食，
至今已有兩千五百多年歷史，
需要經過醃製、烹調等多項工序。

This is an iconic dish of Nanjing,
the ancient capital of China for six dynasties.
This 2,500-year-old recipe calls for multiple steps
including marinating and poaching.

材料

江寧光鴨 _1 隻（1.5 公斤）

花椒鹽料

鹽 _240 克
花椒 _25 克

鹵水香料

八角 _2 顆
草果 _2 個
青花椚（四川紅花椒）_10 克
香葉 _10 片
小茴香 _5 克
白芷 _3 片
薑 _100 克
葱 _100 克
水 _6 公斤
鹽 _160 克
紹興酒 _75 克

鹵水做法

1_ 所有香料放入香料袋內，備用。

2_ 所有鹵水料（紹興酒除外）放入鍋內煮開，以小火煮 1 小時，關火，待用。

綜合做法

1_ 炒鍋不放入油，加入鹽炒熱，下花椒炒勻，關火，盛起花椒鹽，冷卻備用。

2_ 光鴨洗淨，用水沖 2 小時，擦乾水分，將花椒鹽均勻抹在鴨表皮及內腔，冷藏醃 10 小時。

3_ 光鴨取出，洗淨醃料，飛水；將鴨放入鹵水內，用大火燒開，轉小火焗 50 分鐘，加入紹興酒拌勻，用竹籤輕戳入鴨腿肉，見無血水代表熟透，待涼後切件，上碟裝飾。

Ingredients

1 dressed duck from Jiangning District, Nanjing (about 1.5 kg)

Sichuan pepper salt

240 g salt
25 g Sichuan peppercorns

Spiced marinade

2 whole star-anise pods
2 Tsaoko fruits
10 g Sichuan red peppercorns
10 bay leaves
5 g fennel seeds
3 slices Bai Zhi
100 g ginger
100 g spring onion
6 kg water
160 g salt
75 g Shaoxing wine

Method

Spiced marinade

1_ Put all dried herbs and spices into a muslin bag and tie well.

2_ Put all spiced marinade ingredients (except Shaoxing wine) into a pot. Bring to the boil and turn to low heat. Simmer for 1 hour. Turn off the heat.

Poaching duck

1_ Stir-fry salt in a dry wok. Add Sichuan peppercorns. Toss well and turn off the heat. This is Sichuan pepper salt.

2_ Wash the duck and rinse it under a running tap for 2 hours. Wipe dry. Rub Sichuan pepper salt from step 1 evenly over the skin and the inside. Refrigerate for 10 hours.

3_ Take the duck out of the fridge. Rinse off the Sichuan pepper salt. Blanch in boiling water. Drain. Transfer the duck into a pot of spiced marinade. Bring to the boil over high heat. Turn to low heat and simmer for 50 minutes. Add Shaoxing wine and mxi well. Check for doneness by inserting a skewer into the duck leg. It is cooked through if the juices run clear. Remove from the marinade and leave it to cool. Slice and save on a serving plate. Garnish and serve.

HPGC 白切肉

HPGC Plain-blanched Pork

HPGC 白切肉不僅是合萍共廚的經典之作，
亦是淮揚菜的經典菜品。
此前菜源於東北滿族，
後來傳入淮揚地區並得以發揚。

It's not only a signature dish from He Ping Gong Chu,
but also a well-known classic in Huaiyang cuisine.
Though it originated from the Manchurians
in North-eastern China and
was only passed into the Huaiyang region quite a bit later,
it was the Huaiyang chefs who made it famous.

材料

黑毛豬腿肉 _1 塊

醃料

薑片 _30 克
葱 _25 克
鹽 _30 克
油 _30 克

葱蒜汁料

蒜頭 _60 克
小米辣 _2 顆
葱 _100 克
芫荽 _100 克
鮮醬油 _30 克
鹽 _2 克
糖 _5 克
薄鹽生抽 _25 克

做法

1_ 豬腿肉洗淨，擦乾水分，加入薑片、葱及鹽醃勻，再倒入油攪拌醃 90 分鐘。

2_ 蒜頭切末；小米辣切碎，葱切碎，分成葱白及青葱；芫荽切碎，分開梗和葉。

3_ 鍋內放入粟米油，燒熱油至 90℃，放入蒜末微熬，離火，倒入葱白和芫荽梗，盛起放入盤，待涼透後加入鮮醬油、鹽、糖及生抽拌勻，加入青葱及芫荽葉，拌勻成葱蒜汁。

4_ 後腿肉放入蒸鍋蒸 30 分鐘，取出待涼，切成片（每片約 20 克），上碟，淋上葱蒜汁，裝飾即成。

Ingredients

1 leg of Berkshire pig (deboned and skinned)

Marinade

30 g sliced ginger
25 g spring onion
30 g salt
30 g cooking oil

Garlic spring onion sauce

60 g garlic
2 bird's eye chillies
100 g spring onion
100 g coriander
30 g Maggi's seasoning
2 g salt
5 g sugar
25 g low-sodium light soy sauce

Method

1_ Rinse the pork leg. Wipe dry. Add sliced ginger, spring onion and salt. Mix well. Add oil and rub evenly. Marinate for 90 minutes.

2_ Finely chop the garlic. Chop the bird's eye chillies. Finely chop the spring onion and divide into the white parts and the green parts. Finely chop coriander. Separate the leaves and the stems.

3_ To make the garlic spring onion sauce, heat corn oil in a wok up to 90°C. Stir-fry garlic briefly. Turn off the heat. Put in the white parts of spring onion and coriander stem. Mix well and keep in a box. Wait till the mixture is cold. Stir in Maggi's seasoning, salt, sugar and light soy sauce. Mix well. Add green parts of spring onion, and coriander leaves. Mix again.

4_ Steam the pork leg in a steamer for 30 minutes. Leave it to cool. Slice (each slice about 20 g). Arrange on a serving plate. Drizzle with the garlic spring onion sauce. Garnish and serve.

野雞絲

Mock Shredded Pheasant

這是一道傳統名菜，
以生薑絲、南通甜包瓜絲及豬肉絲
三種食材同時入菜，
可謂「鹹甜辣脆、黃金搭配」，
是當地頗具代表性的家常美味。

It's a famous traditional recipe with no pheasant in it –
it's a combination of shredded ginger,
shredded Nantong pickled squash, and shredded pork.
The briny-sweet flavour is nicely balanced by
mild heat of the ginger and crispy texture of the pickled squash.
It's said to be the golden combo of all time,
and a representative delicacy in regional home-made meals.

材料

豬里脊 _120 克
甜包瓜 _240 克
薑 _120g

醃料

花雕酒 _1 克
糖 _1 克
生抽 _3 克
胡椒粉 _ 少許
生粉 _4 克
沙律油 _15 克

調味料

生抽 _3 克
糖 _2 克
蠔油 _3 克

做法

1_ 豬里脊洗淨，切絲，用水泡掉血水，瀝乾水分。

2_ 薑去皮、切絲；甜包瓜去籽，切成細絲，泡水 15 分鐘（期間需換水三次）。

3_ 肉絲加入花雕酒、糖、生抽、胡椒粉及生粉攪拌，再拌入沙律油。

4_ 鍋燒熱，放入沙律油 200 克，待油溫上升至 150℃，放入肉絲迅速用筷子拌散，炒至熟後盛起，備用。

5_ 鍋內放少許油，放入半份薑絲煸炒約 1 分鐘，再放入餘下薑絲及甜包瓜絲，瀝乾水分，加入調味料及肉絲煸炒 1 分鐘，即成。

Ingredients

120 g pork loin
240 g sweet pickled squash
120 g ginger

Marinade

1 g Huadiao wine
1 g sugar
3 g light soy sauce
ground white pepper
4 g caltrop starch
15 g vegetable oil

Seasoning

3 g light soy sauce
2 g sugar
3 g oyster sauce

Method

1_ Rinse the pork and shred it. Soak it in water to drain any blood. Drain well.

2_ Peel and shred the ginger. Set aside. De-seed the pickled squash. Cut into fine strips. Soak them in water for 15 minutes. Drain and change the water every 5 minutes. Repeat 3 times.

3_ The pork mix with marinade ingredients (except vegetable oil). Add vegetable oil at last. Mix again.

4_ Heat a wok and add 200 g of oil. Heat it up to 150°C. Stir-fry the pork quickly with a pair of chopsticks till it's done. Set aside.

5_ Heat a little oil in the same wok. Put in half of the ginger and toss for 1 minute. Add the remaining ginger, and shredded pickled squash. Drain any liquid in the wok. Add seasoning and fried pork from step 4. Toss for 1 minute. Serve.

CHAPTER 2

Freshwater Fish & Shellfish

江鮮及河鮮

Speaking of freshwater fish and shellfish in Huaiyang cuisine, it's mandatory to mention river clams – claimed to be the world's most umami-laden food. Rare species like Chinese tapertail anchovy and puffer fish are often said to be the most representative river fish. On the other hand, critters from the lakes also make excellent dishes. For example, hairy crabs are prized for their rich, buttery roe. In the context of Huaiyang recipes, river clams and Jiangsu-style smoked pomfret are the most famous ones: the former is loaded with umami, whereas the latter is considered an ultimate delicacy in state banquets. Both recipes entail quality ingredients and meticulous craftsmanship.

提及淮揚菜的江鮮與河鮮，不得不提的是被譽為「天下第一鮮」的文蛤，以及珍稀的刀魚與河豚魚，它們構成了江鮮的精髓。河鮮方面，大閘蟹以其肥美著稱。特別值得一提的是，文蛤與蘇式熏鯧魚作為淮揚菜的名菜，前者鮮美無比，後者更是國宴的極品，選料講究，工藝精湛。

富貴河中鮮

Lavish Braised Puffer Fish

顧名思義，其主材料為河鮮，
而富貴一詞來源於選材。
河豚魚作為江浙菜的扛把子，
富含膠原蛋白，口感彈牙，
有「河豚月月鮮」之形容。

Known as the "big boss" of Jiangzhe cooking,
puffer fish is always considered lavish and delectable.
It's rich in collagen, boasting bouncy texture.
Poets compare it to the arrival of spring in April –
full of life and vibrancy.

材料

活河豚魚 _2 條（每條 300 克）

配料

薑片 _20 克

葱 _50 克

油 _50 克

豬油 _25 克

調味料

生抽 _30 克

老抽 _5 克

鹽 _1 克

啤酒 _1.2 公斤

做法

1_ 河豚魚在腹部切開，去腸、剝皮，取出魚肝，洗淨；河豚魚、魚皮及魚肝分別浸泡水中（魚肉泡水 1 小時，以去除血水）。

2_ 燒熱水，放入河豚魚略灼，洗淨，瀝乾水分，備用。

3_ 燒熱水，放入魚皮灼 30 秒，過冷河，去掉魚皮黏液，備用。

4_ 鐵鍋內倒入油及豬油，待油溫至 120℃，放入魚肝炸成淺褐色，盛起。

5_ 放入薑片及葱煸香，離火，放入河豚魚以小火煎 1 分鐘，倒入生抽微煎，再倒入啤酒煮開，灑入鹽，放入魚肝以大火煮開，調至中火燜 25 分鐘，待湯汁煮成半濃稠，放入魚皮煮至湯汁濃稠，加入老抽調色。

6_ 河豚魚肉排放在碟內，魚皮蓋在魚肉上，魚肝擺放魚旁即可。

Ingredients

2 live puffer fish (about 300 g each)

Condiments

20 g sliced ginger
50 g spring onion
50 g cooking oil
25 g lard

Seasoning

30 g light soy sauce
5 g dark soy sauce
1 g salt
1.2 kg beer

Method

1_ Cut open the puffer fish's belly. Remove the guts and save the liver. Peel off the skin and set aside. Rinse the liver well. Soak the fish, the skin and the liver separately in water. Soak the fish in water for at least 1 hour to drain any blood.

2_ Boil water in a pot. Blanch the skinned puffer fish briefly. Rinse and set aside.

3_ Boil water in a pot. Blanch the fish skin for 30 seconds. Rinse in cold water. Scrape off the slime on the skin.

4_ Pour cooking oil and lard into a cast iron wok. Heat it up to 120°C. Deep-fry the liver until lightly browned. Set aside.

5_ In the same wok, put in ginger and spring onion. Fry until fragrant. Turn off the heat. Put in the skinned puffer fish and fry for 1 minute. Add light soy sauce and toss well. Fry for a while longer. Pour in beer and bring to the boil. Season with salt. Add liver and bring to the boil again over high heat. Turn to medium heat and simmer for 25 minutes till the sauce reduces. Add the skin and cook until the sauce is thick. Add dark soy sauce to desired colour.

6. Arrange the skinned puffer fish on a serving plate. Put fish skin over the fish. Put the liver on the side. Serve.

紅燒春刀魚

Red Braised Chinese Tapertail Anchovy

刀魚，江南春饌至鮮，身如銀梭，
脂潤肉細，輕抿即化，脂香漫喉，
宛若匯合一江春水，
盡展江南人對時令風物的虔誠與靈思。

In Jiangnan, Chinese tapertail anchovy is the fattiest and
most delicious in spring.
It looks like a silver spindle; its flesh is fine and oily;
its rich aromas linger in the mouth.
It embodies the essence of life in spring,
as well as the reverence and
ingenuous thoughts of Jiangnan people
towards the transient nature of all things seasonal.

材料

刀魚 _3 條（每條約重 130 克）

配料

肥肉粒 _50 克

薑 _5 片

葱 _5 棵

調味料

生抽 _30 克

糖 _2 克

料酒 _20 克

清雞湯 _100 克

做法

1_ 刀魚去鰓，在魚末端底部處橫劃一小刀，從刀口處拉出魚腸，洗淨備用。

2_ 熱鍋冷油，放入肥肉粒炒至金黃色，放入薑及葱微微煎香，排入刀魚，倒入生抽、料酒及清雞湯，以大火煮開，調至中小火燒至醬汁半濃稠，灑入糖調味，收汁，上碟即成。

Ingredients

3 Chinese tapertail anchovies (about 130 g each)

Condiments

50 g diced fatty pork

5 slices ginger

5 sprigs spring onion

Seasoning

30 g light soy sauce

2 g sugar

20 g cooking wine

100 g chicken stock

Method

1_ Cut off the gills of the fish. Make a small cut on the ventral side of the fish near the tail. Pull out the guts and innards. Rinse well.

2_ Heat a wok and add oil. Stir-fry diced fatty pork until golden. Add ginger and spring onion. Fry until lightly browned. Put in the fish and add light soy sauce, cooking wine and chicken stock. Bring to the boil over high heat. Turn to medium-low heat and simmer until the sauce reduces. Add sugar and reduce the sauce further. Save on a serving plate. Serve.

刀魚魚丸

Chinese Tapertail Anchovy Fish Balls

材料

刀魚 _900 克

配料

甜豆 _20 克

清雞湯 _200 克

薑葱汁 _15 克

粟粉水 _15 克

調味料

鹽 _5 克

糖 _1 克

做法

1_ 刀魚洗淨，瀝乾水分，取肉，去皮，剁成魚肉。

2_ 魚肉放入碗，加入薑葱汁及鹽 3 克，順時針攪拌，下粟粉水融合。

3_ 鍋中加入冷水，用小勺子放入魚丸，開火煮開，調至小火煮 2 分鐘，盛起魚丸備用。

4_ 清雞湯煮開，加入甜豆煮好，灑入鹽及糖調味，下魚丸煮開即成。

Ingredients

900 g Chinese tapertail anchovies

Condiments

20 g sugar snap peas

200 g chicken stock

15 g ginger and spring onion juices

15 g corn starch slurry

Seasoning

5 g salt

1 g sugar

Method

1_ Rinse the fish. Drain well. Debone and skin them. Finely chop the fish meat.

2_ Transfer the fish meat into a bowl. Add ginger and spring onion juices. Sprinkle with 3 g of salt. Stir in clockwise direction until sticky and resilient. Add corn starch slurry and stir until well incorporated.

3_ Pour cold water into a pot. Scoop the fish meat into fish balls and put them in the water. Bring to the boil. Turn to low heat and cook for 2 minutes. Strain and set aside the fish balls.

4_ Boil chicken stock in a pot. Put in sugar snap peas and blanch until cooked through. Season with salt and sugar. Put in the fish balls from step 3. Bring to the boil again. Serve.

文蛤燉蛋

Egg Custard with River Clams

文蛤，在南通自古稱為「天下第一鮮」，
以肉嫩味鮮而著稱。
文蛤燉蛋是淮揚地區家家戶戶經常烹調的菜式，
其滑嫩的口感很受孩子們喜歡。

Since the ancient times, river clams have been prized for their unsurpassed umami and tender texture, known as "the world's most umami-laden food" in Nantong. Egg custard with river clams is a home-style favourite made by every family in the Huaiyang region. Its silky mouthfeel makes it especially popular among children.

材料

文蛤肉 _150 克
雞蛋 _3 個

調味料

鹽 _3 克
水 _300 克
花生油 _10 克
麻油 _5 克

做法

1_ 文蛤肉放入水中順時針攪拌，去掉泥沙，瀝乾水分。

2_ 雞蛋打散，加入鹽、水及花生油，順時針拌勻；倒入碗內，放上文蛤肉，蓋上保鮮紙，蒸 8 分鐘，取出，倒入麻油即可。

Ingredients

150 g shelled river clams

3 eggs

Seasoning

3 g salt

300 g water

10 g peanut oil

5 g sesame oil

Method

1_ Put the shelled clams in a bowl of water and stir clockwise to remove any dirt on them. Drain.

2_ Whisk the eggs. Add salt, water and peanut oil. Stir clockwise until well mixed. Pour into a steaming dish. Arrange shelled clams on top. Cover with cling film. Steam for 8 minutes. Remove from steamer or wok. Drizzle with sesame oil. Serve.

白汁野生鮰魚

White Braised Wild-caught Long-snout Catfish

白汁鮰魚是江蘇揚州、鎮江一帶的特色名菜。
鮰魚學名長吻鮠，
曾被蘇軾、楊慎等多位名家配以詩歌誇讚，
肉質肥嫩鮮美，可見一斑。

This is a famous delicacy from Jiangsu,
Yangzhou and Zhenjiang areas.
Even ancient poets like Su Shi and Yang Shen
wrote verses to celebrate its oily, flavoursome flesh.

材料

鮰魚 _1 條（約 750 克）

配料

高湯 _ 適量（蓋過魚身）
薑片 _30 克
葱 _3 棵
蒜頭 _3 粒
熱豬油 _50 克

調味料

料酒 _30 克
鹽 _8 克
糖 _2 克
胡椒粉 _ 少許

做法

1_ 鮰魚洗淨，沿魚脊骨從頭切至魚尾，使其趴着狀態，放入熱開水微燙，去附黏液，瀝乾水分。

2_ 炒鍋放入熱豬油，加入薑、葱及蒜煸香，鮰魚肉面朝下放入鍋，加入高湯，剛好蓋過魚身為佳，大火燒開，加入料酒、鹽及糖，以中小火煮至收汁，撒上胡椒粉，上碟裝飾即可。

Ingredients

1 long-snout catfish (about 750 g)

Condiments

premium stock (enough to cover the fish)
30 g sliced ginger
3 sprigs spring onion
3 cloves garlic
50 g hot lard

Seasoning

30 g cooking wine
8 g salt
2 g sugar
ground white pepper

Method

1_ Rinse the fish. Pull its backbone from head to tail to make it stand on its belly. Blanch in boiling water briefly and scrape off the slime on its skin. Drain.

2_ Heat up lard in the wok. Fry ginger, spring onion and garlic until fragrant. Put in the fish with the belly side down. Add premium stock to cover the fish. Bring to the boil over high heat. Add cooking wine, salt and sugar. Turn to medium-low heat and cook until the sauce reduces. Sprinkle with ground white pepper. Save on a serving plate and garnish. Serve.

清蒸太湖白條

Steamed Sharpbelly from Lake Taihu

白魚，俗稱白條，是太湖三白（白魚、銀魚、白蝦）之一，
古人早有「白魚如切玉，朱橘不論錢」的詩句讚譽，
是蘇州著名的「太湖船菜」招牌食材，多以清蒸為主。
魚脂化成汁水滲入魚肉，一口吃出肥腴鮮嫩、清甜滑口。

Lake Taihu is famous for three delicacies, namely sharpbelly, anchovy, and white-legged shrimp. Sharpbelly was compared to polished white jade by the Tang Dynasty poet Du Fu, and is a headlining ingredient featured in Taihu Boat Banquet in Suzhou. It is usually steamed so that its fat melts into the flesh to make it juicy, silky, oily ant sweet.

材料

白條魚 _ 約 900 克
熟金華火腿 _8 片（長 6 厘米，寬 3 厘米）
熟竹筍 _8 片（長 6 厘米）
薑 _5 片
葱 _2 棵

醃料

鹽 _3 克
料酒 _15 克

調味料

鮮醬油 _15 克
熱油 _30 克
胡椒粉 _ 少許

做法

1_ 白條魚洗淨，瀝乾水分，放入薑片、葱、鹽及料酒，醃 20 分鐘。

2_ 沖去魚醃料，在魚身輕劃八刀口；魚放於碟內，在每刀口處放上金華火腿及竹筍各一片。

3_ 放入蒸鍋，蒸 7 分鐘後取出，倒入鮮醬油，撒上胡椒粉，最後淋上熱油即可。

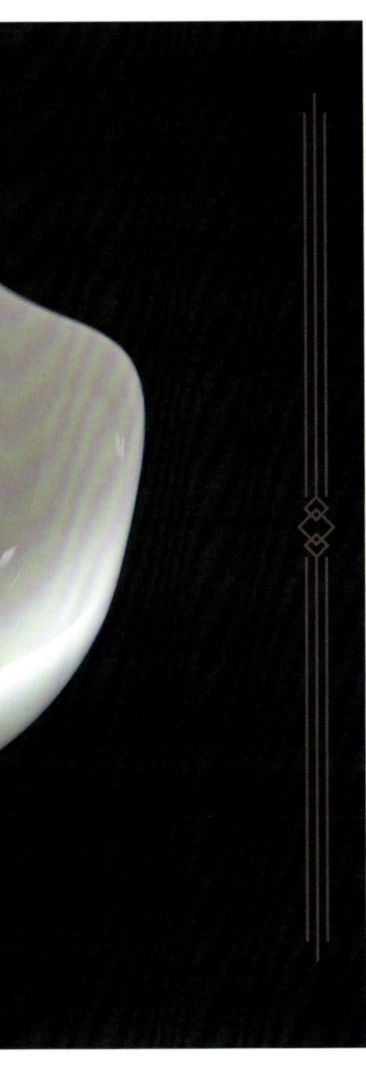

Ingredients

900 g sharpbelly
8 slices cooked Jinhua ham (6 cm by 3 cm each)
8 slices cooked bamboo shoot (6 cm long)
5 slices ginger
2 sprigs spring onion

Marinade

3 g salt
15 g cooking wine

Seasoning

15 g Maggi's seasoning
30 g hot oil
ground white pepper

Method

1_ Rinse the fish and wipe dry. Arrange sliced ginger and spring onion over its skin. Sprinkle with salt and cooking wine. Mix well and leave it for 20 minutes.

2_ Rinse off all marinade. Make 8 light incisions on the fish. Put the fish over a steaming plate and put a slice of Jinhua ham and a slice of bamboo shoot into each incision.

3_ Steam the fish for 7 minutes. Drizzle with Maggi's seasoning and sprinkle with ground white pepper. Lastly drizzle with smoking hot oil. Serve.

薑葱爆河蝦

Stir Fried River Shrimps with Ginger and Spring Onion

淮揚地區豐富的河鮮資源，
為薑葱爆河蝦提供了得天獨厚的食材條件。
一直以來都是尋常人家的待客菜，
做法簡單。

Not only is this a speciality in Huaiyang cooking, but also an indispensable component in the traditional Nantong banquet known as Ba Wan Ba – with four cold appetizers, four stir-fries and eight hot entrées. This recipe is prized for its nutritional value and medicinal properties.

材料

活河蝦 _300 克

配料

生薑 _10 克（切成小片）
葱白段 _10 克
大豆油 _1 公斤
鹽 _2 克

做法

1_ 活河蝦洗淨，瀝乾水分。

2_ 炒鍋放入大豆油，燒至油溫 180℃，倒入河蝦炒勻，炒約 30 秒後，盛起。

3_ 炒鍋再燒熱至 200℃，倒入河蝦，快速盛起，瀝乾油分。

4_ 炒鍋留底油 5 克，放入薑片、葱白段煸香，倒入蝦，均勻地撒上鹽翻炒，上碟即成。

Ingredients

300 g live river shrimps

Condiments

10 g sliced ginger

10 g white part of spring onion (cut into short lengths)

1 litre soybean oil

2 g salt

Method

1_ Rinse the shrimps and wipe dry.

2_ Put soybean oil into a stir-fry wok until 180°C. Put the shrimps in and stir-fry for 30 seconds. Drain and set aside.

3_ Reheat the oil until 200°C. Put the shrimps back and stir-fry quickly. Drain and set aside.

4_ Remain 5 g of the oil in the wok. Stir-fry the ginger and spring onion until fragrant. Put the shrimps in and sprinkle with the salt for the seasoning. Garnish and serve.

梁溪脆鱔

Liang Xi Crispy Eel

這道菜又名無錫脆鱔，是太湖遊船上的船菜，
相傳東漢名人梁鴻曾居於無錫城西一條小河，故名梁溪。
梁溪裏生長一種野生小鱔魚，
烹調成「炸脆鱔」船菜，醬汁帶甜，宜佐茶佐酒。
清末民初，改良後的脆鱔酥脆香口，吃法多樣。

Also known as Wuxi crispy eel,
it is a popular course in the Taihu Boat Banquet.
Legend has it that in Eastern Han Dynasty
a celebrity called Liang Hong once lived on the shore
of a small stream in the western part of Wuxi city.
That's why people called the stream Liang Xi (literally the stream where Liang lives).
There was a breed of wild small eel in the stream
which became the key ingredient for the deep-fried eel course
in Taihu Boat Banquet.
The sauce is sweet, and the dish goes well with teas and wines.
Fast forward to the here and now,
the recipe has evolved over time with many variations.
But still it seduces with crispy texture and rich flavour.

材料

鱔魚肉 _200 克

配料

蒜頭 _5 克
葱段 _5 克
薑 _15 克
生粉 _50 克

調味料

紹興酒 _15 克
糖 _50 克
生抽 _15 克
醋 _15 克
老抽 _5 克

做法

1_ 蒜頭切末；葱切末；薑切絲。

2_ 鱔魚肉斜切成長條，洗淨，用廚房紙吸乾水分。

3_ 鍋內燒熱油至 200℃，鱔魚絲撲上生粉，下油鍋炸至定型，盛起，再下油鍋再炸至金黃酥脆，瀝乾油分。

4_ 鍋內下油 25 克，放入薑、葱及蒜末煸香，加入紹興酒、糖、生抽、醋及老抽煮成糖醋汁，放入炸鱔魚略翻炒，上碟，撒上葱絲點綴即可。

Ingredients

200 g eel (dressed and deboned)

Condiments

5 g garlic

5 g spring onion

15 g ginger

50 g caltrop starch

Seasoning

15 g Shaoxing wine

50 g sugar

15 g light soy sauce

15 g vinegar

5 g dark soy sauce

Method

1_ Finely chop the garlic and spring onion. Finely shred the ginger.

2_ Slice the eel diagonally into long strips. Rinse and wipe dry with paper towel.

3_ Heat oil in a wok up to 200°C. Coat eel strips in caltrop starch. Deep-fry in the hot oil to hold its shape. Remove from oil with a strainer ladle. Deep-fry the eel strips again until golden and crispy. Drain.

4_ Put 25 g of cooking oil in a wok. Stir-fry garlic, spring onion and ginger until fragrant. Add Shaoxing wine, sugar, light soy sauce, vinegar and dark soy sauce. Mix well and toss the eel strips in the sauce. Save on a serving plate. Sprinkle with finely shredded spring onion on top. Serve.

魚米之鄉

The Land Of Plenty
(Sautéed Diced Mandarin Fish with Sweet Corn Kernels)

取骨細肉嫩的桂花魚起肉切丁，
瑩白似玉，
與金黃甜脆的當季粟米粒共舞，
清鮮與甘潤在舌尖交織，
盡顯魚米之鄉的味覺天賜。

Mandarin fish is filleted and diced,
reminiscent of translucent white jade.
Sweet corn in the harvest season
looks golden and tastes crispy.
The two ingredients are woven into
a gustatory tapestry of umami and sweetness,
just like delectable gems from the land of plenty.

材料
桂花魚 350 克

配料
粟米粒 _50 克
青瓜 _30 克
松子仁 _50 克
杞子 _15 克
葱末 _5 克

醃料
蛋清 _1 個
鹽 _4 克
糖 _5 克
胡椒粉 _ 少許
生粉水 _15 克

調味料
花雕酒 _15 克
鹽 _4 克
糖 _5 克

做法

1_ 桂花魚去皮，切成粟米般粒狀，用蛋清、鹽、糖及胡椒粉醃 10 分鐘，拌生粉水上漿；青瓜切成魚粒大小。

2_ 鍋內燒熱油，油溫至 120℃，放入松子仁炒至金黃，盛起；魚米下油鍋拌炒至熟後，盛起。

3_ 鍋內留下油，放入粟米粒、青瓜粒及杞子，灑入花雕酒、鹽、糖及葱末煸熟，再加入魚米、松子仁及生粉水勾獻，略拌即成。

Ingredients

350 g mandarin fish

Condiments

50 g sweet corn kernels
30 g cucumber
50 g pine nuts
15 g goji berries
5 g finely chopped spring onion

Marinade

1 egg white
4 g salt
5 g sugar
ground white pepper
15 g caltrop starch slurry

Seasoning

15 g Huadiao wine
4 g salt
5 g sugar

Method

1_ Dress, fillet and skin the mandarin fish. Dice it into cubes of similar size as sweet corn kernels. Add egg white, salt, sugar and ground white pepper. Mix well and leave it for 10 minutes. Stir in caltrop starch slurry. Set aside. Dice the cucumber into cubes of similar size.

2_ Heat oil in a wok to 120°C. Stir-fry pine nuts until golden. Set aside. Stir-fry the diced fish in the same wok until cooked through. Set aside.

3_ In the same wok, stir-fry sweet corn kernels, diced cucumber and goji berries in the residual oil. Drizzle with Huadiao wine. Sprinkle with salt, sugar and finely chopped spring onion. Toss until cooked through. Put the diced fish back in and add pine nuts. Stir in caltrop starch slurry. Toss until it thickens. Serve.

松鼠鱖魚

Squirrel-shaped Mandarin Fish

此菜相傳是乾隆下江南時賜名，
自此成為宴席頭盤，寓意富貴吉祥。
在魚肉刻上花紋，
裹粉炸至金黃蓬鬆，澆熱糖醋汁，
酸甜開胃，造型靈動。

This iconic dish was believed to be first served to the Qianlong Emperor over 400 years ago. The Emperor loved it and christened it that way because its battered flesh fluffed up like the fur of a squirrel after deep-fried. From then on, it became a popular starter in banquets, symbolizing good fortune and good luck. Diamond-shaped cuts are made on the fish, before it is battered and deep-fried. A sweet and sour sauce is dribbled on top to add a tangy, appetizing touch.

材料

鱖魚 _1 條（約 1.5 公斤）

配料

鳳梨 _60 克

生粉 _60 克

醃料

薑 _20 克

葱 _30 克

花雕酒 _15 克

鹽 _5 克

糖 _2.5 克

糖醋汁

舞茸菇 _30 克

彩椒粒 _ 各 15 克

青豆 _20 克

番茄醬 _80 克

白醋 _15 克

鹽 _5 克

糖 _50 克

生粉水 _35 克

做法

1_ 鱖魚洗淨，在魚鰭處斜切取魚下巴製成松鼠頭；切下整塊魚肉，𠞰上十字花刀，一塊做成松鼠身，另一塊做成松鼠尾巴，加醃料醃 20 分鐘。

2_ 鍋內加入番茄醬、白醋、鹽、糖、舞茸菇、彩椒粒及青豆煮勻，加入生粉水埋獻，調成糖醋汁。

3_ 兩塊魚肉（松鼠身及尾巴）撲上生粉，放入 180℃油鍋炸至定型，盛起，待油溫提升至 220℃，再下魚肉炸至酥脆。魚頭撲上生粉，炸至金黃，盛起。

4_ 鳳梨切厚片，放入平底鍋煎至兩面金黃，上碟，放上炸酥的魚肉及魚頭，淋上糖醋汁即成。

Ingredients
1 mandarin fish (about 1.5 kg)

Condiments
60 g pineapple
60 g caltrop starch

Marinade
20 g ginger
30 g spring onion
15 g Huadiao wine
5 g salt
2.5 g sugar

Sweet and sour sauce
30 g maitake mushrooms
15 g diced yellow bell pepper
15 g diced red bell pepper
20 g green peas
80 g ketchup
15 g white vinegar
5 g salt
50 g sugar
35 g caltrop starch slurry

Method
1_ Rinse the fish. Cut off the head and an angle to include the pectoral fins. This will be the head of the squirrel. Then fillet the fish and make diamond-shaped cuts on the flesh side (while the skin should remain intact). One fillet will be the body of the squirrel and the other will be its tail. Add marinade and mix well. Leave it for 20 minutes.

2_ To make the sweet and sour sauce, heat up ketchup, white vinegar, salt, sugar, maitake mushrooms, bell peppers and green peas. Mix well. Stir in caltrop starch slurry and cook till it thickens.

3_ Coat the fish fillets in caltrop starch. Deep-fry in hot oil at 180°C to hold their shapes. Remove from oil with a strainer ladle. Heat the oil up to 220°C. Deep-fry the fish fillets again until crispy and golden. Set aside. Coat the fish head in caltrop starch. Deep-fry at 220°C until golden. Set aside.

4_ Thickly slice the pineapple. Sear in a pan until both sides golden and caramelized. Arrange on a serving plate. Put the fried fish head and fillets over the pineapple and arrange them like a squirrel. Dribble with sweet and sour sauce on top. Serve.

年糕燜毛蟹

Braised Hairy Crabs with Sticky Rice Cake

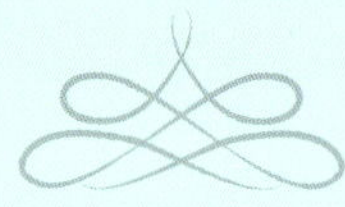

以膏黃豐腴的毛蟹配搭軟糯手打年糕，
濃油赤醬慢煨入味。
糯中帶韌，甜鹹交織，
盡顯通派淮揚菜「以糯鎖鮮」的民間智慧。

Hairy crabs are prized for their rich, buttery roe;
sticky rice cake is soft and chewy in texture;
the reddish brown sauce made with sugar and soy
imbues every bite with deep flavours.
It's a texture and flavour extravaganza that
exemplifies the wisdom of Nantong-style Huaiyang cuisine
which seal in the umami with chewy rice cake.

材料

毛蟹 _4 隻（每隻約 115 克）

年糕 _6 塊

配料

蒜頭 _3 瓣

薑 _15 克

葱 _2 棵

小米辣 _1 隻

生粉 _50 克

調味料

生抽 _20 克

料酒 _20 克

雞湯 _150 克

糖 _3 克

胡椒粉 _0.5 克

香醋 _3 克

麻油 _5 克

做法

1_ 毛蟹洗淨，整隻對切，去掉內臟，備用。

2_ 燒熱油至 180℃，年糕撲上生粉炸至金黃，盛起備用；毛蟹於刀口處撲上生粉炸至金黃（約 1 分鐘），盛起。

3_ 蒜頭、薑、葱及小米辣分別切成小粒，備用。

4_ 鍋內放入少許油，加入薑、蒜及小米辣煸香，放入毛蟹拌勻，下生抽、料酒及雞湯煮 30 分鐘收汁，加入年糕及糖收汁至濃，撒上胡椒粉、香醋及麻油拌炒，最後下葱翻炒，上碟即成。

Ingredients

4 hairy crabs (about 115 g each)
6 pieces sticky rice cake

Condiments

3 cloves garlic
15 g ginger
2 sprigs spring onion
1 bird's eye chilli
50 g caltrop starch

Seasoning

20 g light soy sauce
20 g cooking wine
150 g chicken stock
3 g sugar
0.5 g ground white pepper
3 g vinegar
5 g sesame oil

Method

1_ Rinse the crabs. Cut each in half. Remove the innards.

2_ Heat oil up to 180°C. Coat the sticky rice cake in caltrop starch and deep-fry until golden. Set aside. Coat the cuts of each crab in caltrop starch. Deep-fry for about 1 minute until golden. Set aside.

3_ Finely dice garlic, ginger, spring onion and bird's eye chilli.

4_ Heat some oil in a wok. Stir-fry ginger, garlic and spring onion until fragrant. Add the crabs and toss well. Put in light soy sauce, cooking wine and chicken stock. Cook for 30 minutes. Add the sticky rice cake and sugar. Cook until the sauce reduces. Sprinkle with ground white pepper, vinegar and sesame oil. Toss again. Add spring onion at last. Toss and save on a serving plate. Serve.

蒜子燜鰻魚

Braised Eel with Garlic Cloves

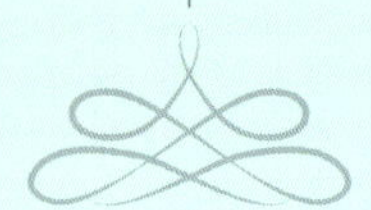

淮揚菜的燜餤一般分為油燜和紅燜兩類，
這道蒜子燜鰻魚採用紅燜的技法，
讓蒜子吃起來並不嗆口，
卻透着軟糯口感。

Braised dishes in Huaiyang cuisine can be categorized into "oil braised" and "red braised" categories.
This recipe is a typical red-braised dish with a reddish brown colour from the soy sauce.
Pungency of the garlic cloves is rounded thanks to prolonged cooking.
They are also soft and creamy like savoury candies.

材料

鰻魚 _750 克

配料

蒜子 _6 瓣

薑片 _15 克

葱 _20 克

調味料

紅燒醬油 _30 克

鹽 _2 克

糖 _15 克

料酒 _30 克

清雞湯 _250 克

做法

1_ 鰻魚洗淨，放入 70℃水略灼，去掉表面黏液；在魚身每隔 1 厘米切一刀，切斷骨但魚肉不切斷。

2_ 炒鍋放入油，放入薑片、葱及蒜子煸香，捲起鰻魚放入鍋微煎，加入紅燒醬油、糖、料酒及清雞湯，煮開後灑入鹽，調至小火，加蓋燜至收汁（約 25 分鐘），上碟即可。

Ingredients
750 g eel

Condiments
6 cloves garlic
15 g sliced ginger
20 g spring onion

Seasoning
30 g soy sauce for red braising
2 g salt
15 g sugar
30 g cooking wine
250 g chicken stock

Method
1_ Rinse the eel. Blanch in water at 70°C briefly. Scrape off the slime on the skin. Make a cut on the eel at 1-inch interval – cut the bones without cutting all the way through.

2_ Heat oil in a wok. Stir-fry ginger, spring onion and garlic cloves until fragrant. Coil the eel and put it into the wok. Fry briefly. Add the soy sauce for braising, sugar, cooking wine and chicken stock. Bring to the boil and sprinkle with salt. Turn to low heat. Cover the lid and cook for about 25 minutes until the sauce reduces. Save on a serving plate. Serve.

鯽魚羊方

Lamb Stuffed with Crucian Carp

最早源於徐州彭祖的「羊方藏魚」，據説已有四千多年歷史。
明清時期廣為流傳，淮揚菜更將其精細化，
融合燉、蒸等技法，滋濃味醇，營養豐富。

This ancient recipe is said to have over 4,000 years of history and can be traced back to the legendary figure Peng Zu. The recipe gained popularity again in Ming and Qing Dynasties when Huaiyang chefs improved it with more finesse and refinement. It entails multiple techniques such as blanching, steaming and stewing. The meaty flavour of lamb synergizes with the umami of the fish for exceptional depth. It's also very nutritious.

材料

鮮羊肉 _1 塊（1.2 公斤）
鮮鯽魚 _1 條（400 克）

醃料

料酒 _10 克
鹽 _5 克
花椒 _1 克
葱段 _10 克
薑片 _10 克

配料

料酒 _50 克
葱段 _20 克
薑片 _20 克
花椒 _1 克
鹽 _5 克
糖 _10 克

做法

1_ 鮮羊肉加入醃料，抹勻羊肉醃 6 小時；放入冷水鍋灼燙，盛起，洗淨。

2_ 鯽魚洗淨，在魚身劃上花刀，灼水後洗淨，抹上鹽 1 克及紹酒 5 克。

3_ 在羊肉側方用刀剖開，放入鯽魚在內。

4_ 將加工的羊肉放入鍋內，倒入清水 2 公斤、鹽 4 克、料酒 50 克、薑片 20 克、花椒及糖，燒開後撇去浮沫，轉小火燉至羊肉酥軟，上碟享用。

Ingredients

1 chunk lamb (1.2 kg)

1 freshly slaughtered crucian crap (400 g)

Marinade

10 g cooking wine

5 g salt

1 g Sichuan peppercorns

10 g spring onion (cut into short lengths)

10 g sliced ginger

Condiments

50 g cooking wine

20 g spring onion (cut into short lengths)

20 g sliced ginger

1 g Sichuan peppercorns

5 g salt

10 g sugar

Method

1_ Rub marinade on the lamb evenly. Leave it for 6 hours. Put it into a pot and add cold water. Bring to the boil. Drain and rinse well.

2_ Rinse the fish. Make diamond-shaped cuts on the fish. Blanch in boiling water. Rinse. Rub 1 g of salt and 5 g of Shaoxing wine on the fish.

3_ Make a cut on the side of the lamb. Stuff the fish into the lamb.

4_ Put the stuffed lamb into a pot. Add 2 kg of water, 4 g of salt, 50 g of cooking wine, 20 g of sliced ginger, Sichuan peppercorns and sugar. Bring to the boil and skim off the foam. Turn to low heat and simmer till the lamb is tender. Serve.

荷包蒸桂花魚

Steamed Mandarin Fish in Chicken Stock Glaze

江南雅饌中的清鮮典範，
魚身瑩白如玉，
脂膏凝潤似琥珀，
輕抿即化，
展現桂花魚的本味之美。

This recipe is an exemplar of
light-tasting delicacies in Jiangnan traditions.
The fish is white and translucent like white jade;
the ham is rich and brightly coloured like amber.
The flaky and melty flesh seduces with natural umami.

材料

桂花魚 _1 條（900 克）
熟金華火腿 _20 片
熟冬筍 _20 片

醃料

薑 _3 片
葱段 _2 棵
料酒 _10 克
鹽 _2 克

獻汁

雞湯 _150 克
鹽 _1 克
糖 _1 克
胡椒粉 _0.5 克
生粉水 _30 克

做法

1_ 桂花魚去骨，取肉，切片，保留魚頭、魚鰭及魚尾。

2_ 魚肉用薑、葱段、鹽及料酒醃 15 分鐘，備用。

3_ 魚頭、魚鰭及魚尾放碟內，將魚片、金華火腿片及冬筍片依次相間地放在魚鰭兩邊，排好。

4_ 放入蒸鍋蒸 4 分鐘，取出，倒掉碟內魚汁。

5_ 燒熱雞湯，放入鹽、糖及胡椒粉拌勻，下生粉水埋獻，淋在魚身上即成。

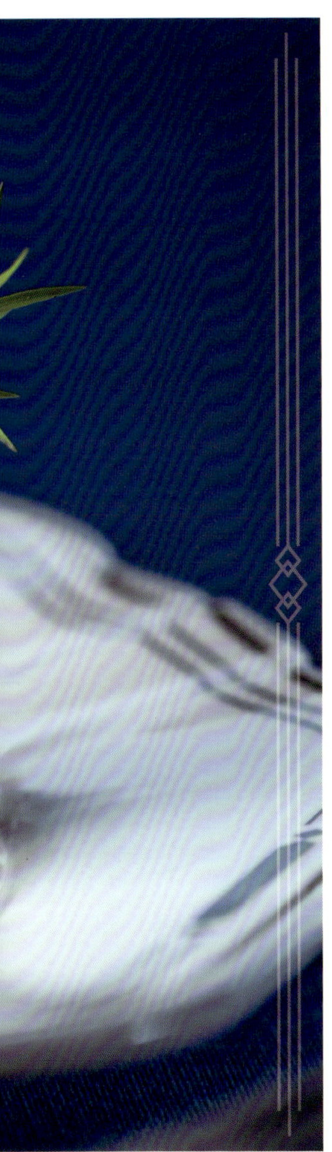

Ingredients

1 mandarin fish (900 g)
20 slices cooked Jinhua ham
20 slices cooked bamboo shoot

Marinade

3 slices ginger
2 sprigs spring onion (cut into short lengths)
10 g cooking wine
2 g salt

Chicken stock glaze

150 g chicken stock
1 g salt
1 g sugar
0.5 g ground white pepper
30 g caltrop starch slurry

Method

1_ Debone, fillet and slice the fish. Keep the fish head, pectoral fins, and tail for later use.

2_ Add ginger, spring onion, salt and cooking wine to the sliced fish. Mix well and leave it for 15 minutes.

3_ In a steaming dish, arrange fish head, pectoral fins and tail along the rim. Then arrange sliced fish, Jinhua ham and bamboo shoot alternately at the centre.

4_ Steam for 4 minutes. Drain any liquid on the dish.

5_ In a small pot, bring chicken stock to the boil. Add salt, sugar and ground white pepper. Mix well. Stir in caltrop starch slurry. Cook till it thickens. Drizzle over the sliced fish. Serve.

CHAPTER 3

Seafood

When it comes to seafood, yellow croaker, big-head croaker, local hairtail, and local pomfret are all representative species of marine fish in Nantong. They may not be especially big in sizes, but they all boast silky flesh and big seafood flavour. You don't need to season them too much, or add too many condiments to bring out their deep umami. They are truly the classics "packing so much umami that your eyebrows fall off after eating them", a hyperbolic vernacular saying in Shanghainese. Though each seafood has its own unique flavour profile, they all impress with exceptional umami.

在南通，對於海鮮而言，黃魚、梅子魚、本灣帶魚及鯧魚都是佼佼者，儘管個頭不大，但肉質十分鮮嫩，在烹調時毋須過多佐料，便可以感受到來自深海的醇厚香氣，是老饕記憶裏「鮮掉眉毛」的經典。各類海鮮以其鮮明的海味個性，撐起了「鮮字當頭」的味覺標杆。

小米遼參

Spiny Sea Cucumber in Millet Porridge

小米遼參是家喻戶曉的名菜，
具減肥、降血脂、降血糖的作用，
製作的重點在於小米在濃湯燉煮時，
必須調至小火烹調。

This is a household name in Huaiyang cuisine
with exceptional medicinal value.
It helps weight loss, reduces blood fats and blood glucose levels.
Just make sure you simmer the millet in the broth over low heat.

材料

乾遼參（70頭）_1支

配料

小米_120克
清雞湯_2公斤
薑_15克
葱_20克
紹興酒_30克
西芹粒_50克

調味料

鹽_5克
糖_2克
胡椒粉_1克
水_500克

做法

1_ 乾遼參放於冷水浸泡 3 天，每隔 12 小時換清水一次，3 天後剪開腹部，去內臟，以小火蒸 1 小時，自然冷卻，再放入冰水泡開（期間換水一次），重複蒸泡步驟至遼參軟硬適中。

2_ 小米洗淨，放入鍋內加入清雞湯，以大火燒開後，調至小火慢煮 30 分鐘。

3_ 海參洗淨，放於熱水鍋，加入薑、葱及紹興酒灼燙。

4_ 海參放入小米粥內，以中小火煮 15 分鐘，加入西芹粒、鹽、糖及胡椒粉調味，上碟品嘗。

Ingredients

1 dried spiny sea cucumber (about 8.6 g)

Condiments

120 g millet
2 kg chicken stock
15 g ginger
20 g spring onion
30 g Shaoxing wine
50 g diced celery

Seasoning

5 g salt
2 g sugar
1 g ground white pepper
500 g water

Method

1_ Soak the dried sea cucumber in cold water for 3 days. Drain and replace with fresh water every 12 hours. Cut open the belly of the sea cucumber. Remove innards and sand. Steam over low heat for 1 hour and leave it to cool in the steamer. Soak it in ice water and leave it in a fridge for 24 hours (drain and replace with fresh water once after 12 hours). Repeat the steaming and ice water soaking steps until the sea cucumber is properly rehydrated.

2_ Rinse the millet and put into a pot. Add chicken stock and bring to the boil over high heat. Turn to low heat and simmer for 30 minutes.

3_ Rinse the sea cucumber. Blanch in boiling water with ginger, spring onion and Shaoxing wine. Drain.

4_ Put the sea cucumber into the millet porridge. Cook over medium-low heat for 15 minutes. Add diced celery, salt, sugar and ground white pepper. Serve.

米湯灼花膠

Fish Maw Blanched in Rice Soup

此菜的特色在於其湯汁濃郁、花膠絲滑，
是一道備受推崇的滋補佳餚。
在製作過程中，
需要注意燉煮的火候與時間。

This dish is prized for the thick, rich soup that contrasts nicely with the velvety texture of the fish maw. Not only is it delicious, but also nourishing and good for health. Pay attention to the heat and the cooking time when you make the rice soup.

材料

花膠（2 頭）_1 片
優質大米 _100 克

配料

藏紅花油 _30 克
鹽 _3 克
糖 _1 克

做法

1_ 乾花膠放於蒸鍋乾蒸，蒸 20 分鐘取出，泡冷水 12 小時。用手指捏花膠厚肉處，至軟硬適中即可，修切成長 10 厘米、寬 4 厘米塊狀。

2_ 大米洗淨，加水 900 克煮開，以小火煮 35 分鐘，再調至中火煮 5 分鐘至濃稠，用手提攪拌機打成糊狀。

3_ 米糊倒入鍋內，拌入藏紅花油，灑入鹽及糖，放入花膠以小火煮至花膠軟糯，上碟即可。

Ingredients

1 dried fish maw (about 300 g)
100 g long-grain rice

Condiments

30 g saffron oil
3 g salt
1 g sugar

Method

1_ Steam the dried fish maw in a steamer for 20 minutes. Soak it in cold water for 12 hours. Pinch the thickest part of the fish maw with your fingers to check if it has rehydrated enough. Cut into strips about 10 cm long and 4 cm wide.

2_ Rinse the rice. Add 900 g of water and bring to the boil. Turn to low heat and simmer for 35 minutes. Turn to medium heat and cook for 5 minutes until thick. Use a hand blender to puree it.

3_ Pour the rice soup into a pot. Add saffron oil, salt and sugar. Put in the fish maw and cook over low heat until the fish maw is soft. Serve.

乾燒大明蝦

Dry-braised Jumbo Shrimps

江海交融的濃鮮盛宴，
湯汁濃稠裹附蝦身，
卻無多餘水分，故名「乾燒」，
其風味鹹鮮微甜，醬香濃郁，
蝦肉緊實彈牙，尾端微焦帶香。

This is an umami-laden feast that
captures the best flavour from the sea and the river.
The dense sauce that clings on the jumbo shrimps
has been reduced so much that it doesn't run.
That's why it's called "dry braised".
Briny sweetness works magic with the strong umami.
The shrimps are firm and bouncy;
their tails are lightly charred, imparting smokiness.

材料

大明蝦 _8 隻
肉末 _50 克

配料

薑 _10 克
葱 _15 克
蒜頭 _3 瓣
小米辣 _1 棵

調味料

番茄醬 _30 克
生抽 _10 克
糖 _10 克
料酒 _15 克
清雞湯 _100 克
鎮江香醋 _5 克
老抽 _3 克

做法

1_ 明蝦洗淨，開背，去腸，備用。

2_ 薑、葱、蒜頭及小米辣切成末，備用。

3_ 燒熱油至 180℃，放入明蝦炸熱，調至 220℃炸至明蝦金黃酥脆，盛起。

4_ 熱鍋冷油，下薑末、蒜末及小米辣煸香，加入番茄醬、生抽、糖、料酒、雞湯及明蝦，以中小火煮至收汁，調入老抽及鎮江香醋拌勻，灑上葱末，上碟即成。

Ingredients
8 jumbo shrimps
50 g ground pork

Condiments
10 g ginger
15 g spring onion
3 cloves garlic
1 bird's eye chilli

Seasoning
30 g ketchup
10 g light soy sauce
10 g sugar
15 g cooking wine
100 g chicken stock
5 g Zhenjiang vinegar
3 g dark soy sauce

Method
1_ Rinse the shrimps. Make a cut along their backs. Devein and set aside.

2_ Finely chop ginger, spring onion, garlic and bird's eye chilli.

3_ Heat oil in a wok up to 180°C. Deep-fry the shrimps to heat through. Turn up the heat to 220°C. Fry until golden and crispy. Drain.

4_ Heat a wok and add oil. Stir-fry ginger, garlic and bird's eye chilli until fragrant. Add ketchup, light soy sauce, sugar, cooking wine, chicken stock and the deep-fried shrimps from step 3. Cook over medium-low heat to reduce the sauce. Stir in dark soy sauce and Zhenjiang vinegar. Mix well and sprinkle with finely chopped spring onion. Serve.

酥香帶魚

Deep-fried Hairtail

作為一道家常菜餚，
酥香帶魚在沿海地區流傳甚廣。
帶魚刺少肉多，
因此在魚類海鮮中很受孩子歡迎。

This is a popular home-style recipe along the coastal regions.
Hairtail is fleshy without many bones.
That explains why it is so popular among children.

材料
本港帶魚（三指寬）_500 克

配料
薑片 _15 克
葱段 _50 克
蒜頭 _10 克
生粉 _150 克
鎮江香醋 _20 克

醃料
鹽 _5 克
胡椒粉 _1 克
料酒 _15 克

醬汁
生抽 _10 克
料酒 _15 克
糖 _28 克
花椒油 _4 克
鎮江香醋 _20 克

做法

1_ 帶魚洗淨，切成約 7 厘米長段，用醃料拌勻醃 30 分鐘，用廚房紙吸乾水分，撲上生粉。

2_ 將汁料調成醬汁，備用。

3_ 煎鍋放入油，加入薑片、葱段及蒜片煸香，油溫調至 120℃，下帶魚煎至兩面金黃，每面各煎 1 分鐘。

4_ 調入醬汁，調至小火，讓帶魚兩面裹上汁，最後倒入鎮江香醋 20 克，微㸆收汁，裝盤點綴即可。

Ingredients

500 g local hairtail (about the width of three fingers)

Condiments

15 g sliced ginger

50 g spring onion (cut into short lengths)

10 g garlic

150 g caltrop starch

20 g Zhenjiang vinegar

Marinade

5 g salt

1 g ground white pepper

15 g cooking wine

Sauce

10 g light soy sauce

15 g cooking wine

28 g sugar

4 g Sichuan pepper oil

20 g Zhenjiang vinegar

Method

1_ Rinse the fish. Cut into segments about 7 cm long. Add marinade and mix well. Leave it for 30 minutes. Wipe dry with paper towel. Coat it in caltrop starch.

2_ Mix the sauce ingredients together.

3_ Heat oil in a wok. Fry ginger, spring onion and garlic until fragrant. Bring oil temperature to 120°C. Fry the fish for 1 minute on each side until both sides golden.

4_ Pour in the mixed sauce. Turn to low heat. Toss the fish for the sauce to cling on both sides. Drizzle with 20 g of Zhenjiang vinegar at last. Cook briefly to reduce the sauce. Garnish and serve.

香酥梅子魚

Deep-fried Big-head Croaker

南通人稱之為梅子魚，
江浙地區有「冷水梅童賽黃魚」之説。
香酥梅子魚的特色在於外皮酥脆、魚肉鮮嫩，
烹飪過程中需經過醃製、裹粉及炸製等工序，
才能做出完美口感。

This is how an old saying in Jiangzhe region goes,
“Big-head croaker is comparable to yellow croaker in winter.”
This speciality is characterised by
its crispy skin and succulent flesh.
For the perfect result, pay attention to the marinating,
battering and deep-frying steps.

材料

梅子魚 _400 克

配料

青椒粒 _1 克

紅椒粒 _1 克

椒鹽 _2 克

醃料

葱 _20 克

薑 _15 克

鹽 _2 克

料酒 _15 克

蛋黃糊

蛋黃 _6 個

麵粉 _80 克

芝士粉 _20 克

水 _40 克

鹽 _2 克

做法

1_ 梅子魚洗淨，瀝乾水分，放入薑、葱、鹽及料酒醃 20 分鐘，以吸油紙擦乾魚身表面。

2_ 蛋黃糊材料攪拌均匀，放入梅子魚蘸匀，備用。

3_ 放入油燒至 160℃，以筷子夾着魚尾放入油鍋炸 2 分鐘，盛起，油溫調至 180℃，再放入魚炸 30 秒，瀝乾油分。

4_ 另起鍋，放入油 5 克，下青紅椒粒煸香，放入魚拌匀，撒上椒鹽翻炒，上碟點綴即可。

Ingredients
400 g big-head croaker

Condiments
1 g diced green bell pepper
1 g diced red bell pepper
2 g peppered salt

Marinade
20 g spring onion
15 g ginger
2 g salt
15 g cooking wine

Deep-frying batter
6 egg yolks
80 g flour
20 g grated parmesan
40 g water
2 g salt

Method
1_ Rinse the fish and drain well. Add the marinade and mix well. Leave it for 20 minutes. Wipe dry the skin with paper towel.

2_ Make the deep-frying batter by mixing all ingredients together until lump-free. Dip the fish into the batter.

3_ Heat oil in a wok up to 160°C. Hold the fish by its tail with a pair of chopsticks. Dip it into the oil with the head going in first for 2 minutes. Remove from hot oil. Heat the oil up to 180°C. Put in the fish to fry for 30 seconds more. Drain.

4_ In another wok, put in 5 g of cooking oil. Stir-fry green and red bell pepper until fragrant. Put in the fish and toss well. Sprinkle with peppered salt and toss again. Save on a serving plate and garnish. Serve.

金湯黃魚脯

Yellow Croaker Fillet in Golden Broth

金湯黃魚脯源於古代皇家宴席，
選用優質大黃魚為主料，
可搭配蛤蜊、南瓜等多種食材，
其特色在於湯汁金黃濃郁。

This recipe can be traced back to the imperial banquets in ancient China. Giant yellow croaker is paired with a broth that is rich in flavour and golden in colour. For variations, you may add clams or pumpkin.

材料

大黃魚 _750 克

醃料

鹽 _3 克
薑 _15 克
葱 _15 克
料酒 _15 克

調味料

濃雞湯 _300 克
鹽 _2 克
糖 _2 克
雞油 _20 克
藏紅花油 _20 克
生粉水 _30 克

做法

1_ 黃魚洗淨，開背去龍骨，放入醃料醃 30 分鐘。

2_ 黃魚切成塊狀，放於蒸碟內蒸 8 分鐘，倒出盤內多餘的汁。

3_ 濃雞湯煮開，放入鹽、糖及雞油拌勻，下生粉水勾玻璃獻，推入藏紅花油至均勻，淋在魚身上即可。

Ingredients
750 g giant yellow croaker

Marinade
3 g salt
15 g ginger
15 g spring onion
15 g cooking wine

Seasoning
300 g condensed chicken stock
2 g salt
2 g sugar
20 g chicken fat
20 g saffron oil
30 g caltrop starch slurry

Method
1_ Rinse the fish and cut along the back to fillet it. Add marinade and mix well. Leave it for 30 minutes.

2_ Cut the fish fillet into slices. Arrange on a steaming plate. Steam for 8 minutes. Drain any liquid on the plate.

3_ Bring condensed chicken stock to the boil. Add salt, sugar and chicken fat. Stir in caltrop starch slurry and cook until it thickens slightly. Stir in saffron oill and mix well. Dribble over the steamed fish fillet from step 2. Serve.

蝦仁鍋巴

Shrimps on Scorched Rice

又稱為「平地一聲驚雷」、以「金玉交響」聞席，
用蝦仁及鍋巴為主料製成，
也是江南地區一道傳統名菜。

Nicknamed "a clap of thunder"
for the snapping sound the scorched rice makes,
as well as "the world's number one dish"
for the contrasting textures and deep flavours,
this is an exemplar in Jiangnan culinary traditions.

材料

蝦仁 _100 克
乾鍋巴 _10 片

漿料

鹽 _10 克
糖 _30 克
蛋清 _1 個
生粉 _5 克

配料

青豆 _50 克
番茄醬 _30 克
白醋 _20 克
清雞湯 _300 克
胡椒粉 _ 少許

做法

1_ 蝦仁洗淨，瀝乾水分，加入鹽、糖、蛋清及生粉拌成漿，放入油鍋輕燙成型。

2_ 鍋留底油，下番茄醬及白醋炒至有色，加入蝦仁、青豆、清雞湯及胡椒粉調味，燒沸後下生粉水勾獻，盛入碗內。

3_ 鍋燒油至七成熱，下鍋巴炸至金黃酥脆，置於碟內，與蝦仁汁同時上桌，享用時將蝦仁汁淋在鍋巴即可。

Ingredients
100 g shelled shrimps
10 pieces dried scorched rice

Batter
10 g salt
30 g sugar
1 egg white
5 g caltrop starch

Condiments
50 g green peas
30 g ketchup
20 g white vinegar
300 g chicken stock
ground white pepper

Method
1_ Rinse the shrimps and wipe dry. Add salt, sugar, egg white and caltrop starch. Mix well and coat the shrimp evenly. Shallow-fry in oil briefly to hold their shapes.

2_ In the same wok, add ketchup and white vinegar to the remaining oil. Toss until the mixture turns darker in colour. Put in the shrimps, green peas, chicken stock and ground white pepper. Bring to the boil and stir in the caltrop starch slurry. Cook till it thickens. Save in a bowl.

3_ Heat oil in a wok up to 180°C. Deep-fry the scorched rice until golden and crispy. Save on a serving plate. Serve with the shrimp glaze on the side. Pour the shrimp glaze over the scorched rice right before eating.

蘇式鯧魚

Jiangsu-style Smoked Pomfret

燻魚是淮揚菜系傳統的風味名吃。
鯧魚作為中國沿海一帶的特產，
為淮揚菜的重要食材之一，
蘇式鯧魚在這個文化背景中孕育而生。

Smoked fish is a famous recipe in Huaiyang cuisine,
but again, is a misnomer.
The fish is not really smoked;
but deep-fried and steeped in a flavoursome marinade.
For this recipe, pomfret is used because
it is one of the representative ingredients in Huaiyang cooking,
available along China's coastal areas,
a cradle of Chinese civilisation.
This recipe is born to the rich cultural background of
the transportation thoroughfare.

材料

鯧魚 _2 條（每條 200 克）

醃料

薑片 _15 克
葱 _25 克
鹽 _3 克
料酒 _15 克

燻魚汁

八角 _3 顆
桂皮 _5 克
香葉 _5 片
薑片 _15 克
萬字醬油 _18 克
草菇老抽 _5 克
鎮江香醋 _18 克
米酒 _20 克
紹興酒 _25 克
蠔油 _20 克
海鮮醬 _50 克
蜂蜜 _11 克
冰糖 _90 克
鹽 _2 克
水 _500 克

做法

1_ 鯧魚洗淨，擦乾水分，加入薑、葱、鹽及料酒醃 30 分鐘，用廚房紙吸乾水分。

2_ 熏魚汁放入鍋內煮至濃稠（約 30 分鐘），盛起。

3_ 開油鍋燒至 160℃，放入鯧魚炸 2 分鐘，盛起，待油溫調升至 190℃，下鯧魚再炸約 1 分鐘至表面硬脆，盛起，淋上熏魚汁，點綴即可。

Ingredients
2 pomfrets (about 200 g each)

Marinade
15 g sliced ginger
25 g spring onion
3 g salt
15 g cooking wine

Smoked fish sauce
3 cloves star-anise
5 g cassia bark
5 bay leaves
15 g sliced ginger
18 g Kikkoman soy sauce
5 g mushroom-flavoured dark soy sauce
18 g Zhenjiang vinegar
20 g rice wine
25 g Shaoxing wine
20 g oyster sauce
50 g Hoi Sin sauce
11 g honey
90 g rock sugar
2 g salt
500 g water

Method
1_ Rinse the fish and wipe dry. Add the marinade and mix well. Leave them for 30 minutes. Wipe dry with paper towel.

2_ To make the smoked fish sauce, put all ingredients into a pot. Bring to the boil and cook for about 30 minutes until thick and sticky. Set aside.

3_ Heat oil in a wok up to 160°C. Deep-fry the fish for 2 minutes. Remove from hot oil. Turn the heat up and wait till the oil temperature reaches 190°C. Put in the fish to fry for 1 minute more until crispy. Save on a serving plate. Pour on the sauce. Garnish and serve.

大烤墨魚

Braised Cuttlefish

這是一道經典的沿海風味菜，
墨魚肉質緊實有嚼勁，
帶有海鮮的天然鮮味，
佐飯或下酒皆宜，
盡顯沿海飲食對「鮮」與「味」的巧妙融合。

This is a culinary classic along the coastal region.
Cuttlefish is firm in texture and loaded with natural umami,
making it the perfect companion to
alcoholic drinks or steamed rice.
It also embodies the wisdom of melding umami with
flavours among coastal dwellers.

材料

新鮮墨魚 _1 隻（2.1 公斤）

配料

薑 _15 克

葱 _10 克

八角 _2 顆

香葉 _1 片

調味料

南乳汁 _80 克

生抽 _10 克

老抽 _10 克

辣鮮露 _15 克

糖 _5 克

鹽 _5 克

花雕酒 _15 克

蠔油 _8 克

清水 _2 公斤

做法

1_ 新鮮大墨魚撕去外皮，去內臟，洗淨。

2_ 鍋內放入清水，燒開後放入墨魚飛水，盛起備用。

3_ 燒熱少許油，加入薑、葱、八角及香葉煸香，放入墨魚、清水及所有調味料，大火燒開後撇去浮沫，轉小火加蓋燜煮 45 分鐘。

4_ 墨魚煮至軟身，調至大火收汁，盛起以生菜墊底，整成條狀，上碟。

Ingredients

1 fresh cuttlefish (about 2.1 kg)

Condiments

15 g ginger
10 g spring onion
2 cloves star-anise
1 bay leaf

Seasoning

80 g pickling brine of red tarocurd
10 g light soy sauce
10 g dark soy sauce
15 g spicy Maggi's seasoning
5 g sugar
5 g salt
15 g Huadiao wine
8 g oyster sauce
2 kg water

Method

1_ Peel off the skin of the cuttlefish. Remove the innards. Rinse well.

2_ Boil some water in a pot. Blanch the cuttlefish. Drain.

3_ Heat some oil in a wok. Stir-fry ginger, spring onion, star-anise and bay leaves until fragrant. Add cuttlefish and water. Then put in all remaining seasoning. Bring to the boil over high heat. Skim off the foam on the surface. Turn to low heat and cover the lid. Simmer for 45 minutes.

4_ Check if the cuttlefish has turned tender. Turn to high heat to reduce to sauce. Line the bottom of a dish with lettuce leaves. Cut the cuttlefish into strips. Arrange over the bed of lettuce leaves. Serve.

絲瓜文蛤蟶乾湯

River Clam and Dried Razor Clam Soup with Angled Loofah

這是南通地區夏季常見的家常湯品，
充滿沿海風味。
以鮮嫩絲瓜搭配肥美的文蛤及蟶子，
湯汁清鮮甘潤，
盡顯江海食材的鮮甜。

A popular home-style soup in summer for Nantong residents, it is loaded with seafood sweetness and umami. Crisp and refreshing angled loofah is paired with plump river clams and dried razor clams for a nourishing starter to beat the summer heat.

材料

水發螺乾 _150 克
文蛤肉 _100 克
絲瓜 _1 個

配料

基圍蝦 5 隻
金華火腿 30 克
雞蛋 _1 個
春筍 _1/2 個
浸發木耳 _15 克
薑絲 _15 克
葱段 _10 克

調味料

高湯 _350 克
料酒 _15 克
糖 _2 克
鹽 _5 克
胡椒粉 _0.5 克

做法

1_ 雞蛋拂打，煎成蛋皮，切粗絲。

2_ 春筍切片，灼水 5 分鐘，瀝乾水分。

3_ 絲瓜去皮，切成滾刀塊，泡鹽水備用。

4_ 金華火腿泡水 20 分鐘，蒸 10 分鐘，取出待涼，切薄片。

5_ 基圍蝦挑去腸，放入滾水內，關火待 1 分鐘，泡冰水待涼，去殼取肉備用。

6_ 鍋內下油，放入薑絲及葱段煸香，下螺乾及高湯，放入春筍、木耳及絲瓜，煮開後下料酒拌勻煮滾，放入文蛤肉及其他調味料，最後撒入蛋絲，盛起即可。

Ingredients

150 g rehydrated dried razor clams
100 g shelled river clams
1 angled loofah

Condiments

5 shrimps
30 g Jinhua ham
1 egg
1/2 bamboo shoot
15 g wood ear fungus (rehydrated)
15 g shredded ginger
10 g spring onion (cut into short lengths)

Seasoning

350 g premium stock
15 g cooking wine
2 g sugar
5 g salt
1/2 g ground white pepper

Method

1_ Whisk the egg. Fry in a pan with oil to make a thin omelette. Cut into thick strips.

2_ Slice the bamboo shoot. Blanch in boiling water for 5 minutes. Drain.

3_ Peel angled loofah. Cut into random wedges while rolling the angled loofah slowly on a chopping board. Soak in salted water. Drain right before use.

4_ Soak Jinhua ham in water for 20 minutes. Steam 10 minutes, then let it cool. Thinly slice the ham.

5_ Remove the intestine of shrimps. Put the shrimps in the boiling water, remove from heat and keep in the boiling water for 1 minute. Soak in the iced water. Shell them and pick the meat. Set aside.

6_ Heat oil in a wok and stir-fry shredded ginger and spring onion until fragrant. Add dried razor clams and premium stock. Toss and add bamboo shoot, wood ear fungus and angled loofah. Bring to the boil and add cooking wine. Stir well and bring to the boil. Add the river clams and all remaining seasoning. Sprinkle with omelette strips on top. Serve.

CHAPTER 4

Meat & Poultry

禽肉

I grew up in an underprivileged rural community. In an era of material deprivation, meat was a luxury and having meat on the table was undoubtedly the happiest moment in my childhood life. At an age of eleven, I was particularly fond of red-braised pork belly and my mother loved me to bits. As I was in the stage of growth spurts, she thoughtfully made over 300 grams of red-braised pork belly for me every day, and paired it with Tatsoi, a classic leafy green in Huaiyang cooking. I still remember the dish vividly even now. Both braised pork and chicken are home-style staples in Huaiyang cuisine, highlighting the natural flavours of the ingredients without overpowering seasoning. These dishes always give me comfort and the peace of mind, and their flavours are simply unforgettable.

我成長於貧寒的農村，在那個物質匱乏的年代，能吃到肉食無疑是一天中最幸福的時刻。十一歲那年，我尤其鍾愛紅燒肉，媽媽心疼我正處於發育期，會每天精心烹製半斤以上的紅燒肉，並搭配淮揚菜的經典塔菜，讓我記憶猶新。

談及淮揚菜中的扣雞、扣肉，它們均為地道的家庭美食，生動還原了當地肉類、禽類的原汁原味，讓人吃的放心，回味無窮。

上海八寶雞

Shanghainese Eight-treasure Chicken

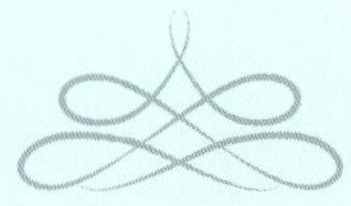

一道融合了上海本地風味與江南特色的佳餚，
全雞出骨，八寶填料，蒸製與收汁，
肉質酥爛，八寶香氣滲透，
口感層次豐富，香濃可口。

This recipe melds Shanghainese flavours with Jiangnan characteristics.
The chicken is deboned and stuffed with eight-treasure filling, before steamed and drizzled in a reduced sauce.
The chicken will end up tender and melty; the filling richly perfumed, delivering layers of textures and big flavours.

材料

母雞 _1 隻（1 公斤）
荷葉 _1 張

配料

糯米 _80 克
江瑤柱 _10 克
冬筍 _250 克
水發香菇 _2 朵
金華火腿 _30 克
小青豆 _20 克
浸發海參 _20 克
紅蘿蔔 _30 克

調味料

葱 _15 克
薑 _20 克
鹽 _1 克
糖 _8 克
生抽 _10 克
蠔油 _15 克
料酒 _15 克
胡椒粉 _ 少許

做法

1_ 母雞去內臟，洗淨，從頸項與雞身交接處下刀，去骨（注意下刀時緩慢，雞皮不要弄破），加入調味料醃 20 分鐘。

2_ 糯米浸泡一晚，瀝幹水分，放入蒸籠蒸 30 分鐘。

3_ 江瑤柱、冬筍、香菇、金華火腿、海參及紅蘿蔔切成丁，和小青豆飛水，瀝乾水分。

4_ 配料與糯米拌勻，加入鹽、糖、胡椒粉、生抽及蠔油拌成八寶料。

5_ 將八寶料填滿雞腹，不能太滿，用泡軟的荷葉包好，放入蒸籠蒸 30 分鐘即成。

Ingredients
1 dressed hen (1 kg)
1 lotus leaf

Condiments
80 g glutinous rice
10 g dried scallops
250 g bamboo shoot
2 dried shiitake mushrooms (rehydrated)
30 g Jinhua ham
20 g green peas
20 g sea cucumber (rehydrated)
30 g carrot

Seasoning
15 g spring onion
20 g ginger
1 g salt
8 g sugar
10 g light soy sauce
15 g oyster sauce
15 g cooking wine
ground white pepper

Method
1_ Remove the innards from the hen. Rinse well. Make a cut where the chicken neck connects to the body. Debone the chicken. Make sure you do it slowly without breaking the skin. Add seasoning ingredients. Mix well and leave it for 20 minutes.

2_ Soak glutinous rice in water overnight. Drain and transfer into a bamboo steamer. Steam for 30 minutes.

3_ Dice dried scallops, bamboo shoot, shiitake mushrooms, Jinhua ham, sea cucumber and carrot. Blanch them in boiling water together with green peas. Drain.

4_ Mix all blanched ingredients from step 3 with glutinous rice. Add salt, sugar, ground white pepper, light soy sauce and oyster sauce. This is the eight-treasure filling.

5_ Stuff the chicken with the filling from step 4 until 70% full. Wrap the chicken in a lotus leaf that has been soaked in water till soft. Steam in a bamboo steamer for 30 minutes. Serve.

冰糖蹄膀

Braised Pork Knuckle in Rock Sugar Soy Sauce

江南地區傳統的經典名菜，其製作過程較為講究，
成菜後，冰糖蹄膀色澤紅亮，
肉質酥爛脱骨，肥而不膩，
皮層軟糯黏嘴，甜鹹交融，口感豐富。

This Jiangnan classic is a bit finicky to make.
But once you succeed,
you'd end up with a pork knuckle in shiny red colour,
with melty flesh that falls off the bones,
and sticky, gelatinous skin.
The sweetness and saltiness are rounded and mellow,
and the exquisite mouthfeel is pure delight.

材料

豬前蹄膀 _1 隻（1 公斤）
豆苗 _200 克

配料

葱 _10 克
薑片 _10 克
紹興酒 _100 克
醬油 _50 克
鹽 _3 克
冰糖 _100 克
油 _80 克

做法

1_ 豬蹄膀洗淨，放於爐火烤至外皮微黃。

2_ 豬蹄膀放入盛有水的鍋內，飛水 5 分鐘至外皮收縮，取出洗淨。

3_ 鍋內放入油 40 克，下冰糖 30 克煮至出現糖色，盛起。

4_ 鍋內放入豬蹄膀、冰糖漿、醬油及水，蓋過豬蹄，放入其餘配料（紹興酒除外）煮，倒入紹興酒煮 90 分鐘，測試肉質軟腍，以大火收汁，上碟。

5_ 鍋內放入油 40 克，加入豆苗爆炒，灑入鹽 2 克炒勻，圍在豬蹄膀即可。

Ingredients

1 pork knuckle (forelimb, about 1 kg)
200 g pea tips

Condiments

10 g spring onion
10 g sliced ginger
100 g Shaoxing wine
50 g soy sauce for red braising
3 g salt
100 g rock sugar
80 g cooking oil

Method

1_ Rinse the pork knuckle. Grill over a stove until the skin is lightly browned.

2_ Blanch the pork knuckle in boiling water for 5 minutes until the skin shrinks. Rinse and drain.

3_ Put 40 g of oil in a wok. Add 30 g of rock sugar. Cook until the sugar is caramelised. Set aside this syrup.

4_ In the same wok, put in the pork knuckle, the syrup from step 3 and the remaining rock sugar. Add enough water to cover. Then put in the remaining condiments (except Shaoxing wine). Bring to the boil. Add Shaoxing wine. Cook for 90 minutes. Check the consistency by inserting a chopstick into the fleshiest part of the pork knuckle. Cook further if necessary. Turn up the heat to reduce the sauce. Save on a serving dish.

5_ Pour 40 g of cooking oil in a wok. Stir-fry pea tips until they wilt. Sprinkle with 2 g of salt and toss well. Arrange around the pork knuckle on the serving dish. Serve.

紅燒海門羊肉

Red Braised Haimen Goat

海門區隸屬南通市，其羊肉以質地鮮嫩、無羶味而聞名，
是一道具有濃郁地方特色的傳統名菜，
海門羊肉的烹製技藝更入選第五批江蘇省非遺專案名錄。

Haimen is a district in the city of Nantong in Jiangsu Province.
It is famous for its goat that
boasts tender, silky flesh without gamey taste.
This recipe is a regional speciality deeply rooted in traditions.
The techniques used to prepare Haimen goat even made it to
the fifth cohort of intangible cultural heritage of Jiangsu Province.

材料

海門山羊肉 _1 公斤

配料

薑片 _40 克
葱 _40 克
紅燒醬油 _30 克
料酒 _100 克
冰糖 _18 克
開水 _1.5 公斤

做法

1_ 羊肉切成長及寬 5 厘米大塊，放於冷水沖 1 小時，去掉血水。

2_ 羊肉放入冷水下鍋，加入薑片 20 克、葱 20 克及料酒 30 克，以中火煮開，撇去浮沫，盛起羊肉，沖水，瀝幹水分。

3_ 鍋內放入油，加入餘下薑片及葱煸香，放入羊肉炒出香味，倒入紅燒醬油及冰糖，炒至冰糖溶化，放入料酒 70 克，加入開水煮滾，改中小火加蓋燜 50 分鐘至羊肉酥軟。

4_ 最後，大火收汁至濃稠，上碟裝飾即可。

Ingredients

1 kg Haimen goat meat

Condiments

40 g sliced ginger
40 g spring onion
30 g soy sauce for red braising
100 g cooking wine
18 g rock sugar
1.5 kg boiling hot water

Method

1_ Cut goat meat into chunks, about 5 cm by 5 cm. Rinse in cold water for 1 hour to remove any blood.

2_ Pour cold water into a pot. Put in the goat meat. Add 20 g of ginger, 20 g of spring onion and 30 g of cooking wine. Bring to the boil over medium heat. Skim off the foam on the surface. Drain and rinse the goat meat. Drain again.

3_ Heat oil in a wok and stir-fry the remaining ginger and spring onion. Add goat meat and toss until fragrant. Add soy sauce for red braising and rock sugar. Cook until the sugar dissolves. Add 70 g of cooking wine and boiling water. Bring to the boil again. Turn to medium-low heat and cover the lid. Simmer for 50 minutes until the goat meat is tender.

4_ Turn the heat up to reduce the sauce. Save on a serving dish and garnish. Serve.

醃篤鮮

Yanduxian
(Cured Pork and Fresh Pork Soup with Bamboo Shoot and Tofu Skin Knots)

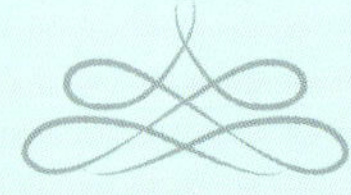

醃篤鮮為江南一道傳統名菜，
蘇州本地人往往會以「鮮得掉眉毛」來形容，
其重點在於小火慢燉，講究一個「鮮」字。

This Jiangnan classic broth is commonly described by Suzhou natives as "packing so much umami that your eyebrows fall out" after eating it.
Of course, it's just a hyperbolic figure of speech.
The key technique lies in the use of low heat to slow cook the broth,
bringing out the ultimate umami in the ingredients.

材料

鹹肉 _100 克
帶皮五花肉 _150 克
春筍 _100 克
百葉結 _6 個
萵筍 _5 塊

調味料

鹽 _2 克
糖 _1 克

配料

薑片 _30 克
葱 _15 克
清雞湯 _750 克
紹興酒 _15 克
食用梳打粉 _1 克

做法

1_ 鹹肉切成塊狀（長 4 厘米、寬 3 厘米、厚 2 厘米），浸泡於水 20 分鐘。

2_ 五花肉切成長條塊（長 5 厘米、寬 4 厘米、厚 3 厘米），浸泡鹽水 20 分鐘。

3_ 水 250 克與食用梳打粉 1 克拌勻，放入百葉結浸泡 15 分鐘，取出，沖洗乾淨。

4_ 春筍去皮，切滾刀塊，放入盛有水 750 克及糖 15 克的鍋內，用中火煮 10 分鐘，沖洗乾淨，瀝乾水分。

5_ 鹹肉及五花肉分別飛水，洗淨，瀝乾水分，備用。

6_ 湯鍋內倒入清雞湯，放入薑片、葱、鹹肉及五花肉煮開，倒入紹興酒用大火煮滾，調至小火煮 90 分鐘，加入春筍、百葉結及萵筍再煮，最後灑入鹽及糖煮 5 分鐘，即可。

Ingredients

100 g cured pork
150 g skin-on pork belly
100 g bamboo shoot
6 tofu skin knots
5 pieces celtuce

Condiments

30 g sliced ginger
15 g spring onion
750 g chicken stock
15 g Shaoxing wine
1 g baking soda

Seasoning

2 g salt
1 g sugar

Method

1_ Cut cured pork into rectangular pieces (about 4 cm by 3 cm each, 2 cm thick). Soak in water for 20 minutes.

2_ Cut pork belly into strips (about 5 cm by 4 cm each, 3 cm thick). Soak in salted water for 20 minutes.

3_ In a bowl, add 250 g of water and 1 g baking soda. Mix well. Soak the tofu skin knots in the baking soda solution for 15 minutes. Drain and rinse well.

4_ Peel the bamboo shoot. Cut into random wedges while rolling it slowly on a chopping board. Put 750 g of water and 15 g of sugar into a pot. Add bamboo shoot. Bring to the boil and turn to medium heat. Cook for 10 minutes. Rinse and drain.

5_ Blanch cured pork and pork belly separately in boiling water. Rinse and drain.

6_ In a soup pot, pour in chicken stock. Add ginger, spring onion, cured pork and pork belly. Bring to the boil. Add Shaoxing wine and bring to the boil over high heat. Turn to low heat and simmer for 90 minutes. Add bamboo shoot, tofu skin knots and celtuce. Bring to the boil again. Sprinkle with salt and sugar. Cook for 5 minutes. Serve.

八寶雞翅

Eight-treasure Stuffed Chicken Wings

此菜名為「八寶雞翅」，
因為廚師們將傳統的八寶餡料（如蓮子、芡實、糯米……）
巧妙地釀入雞翅當中，
是典型的淮揚菜清鮮平和的口味。

Eight traditional ingredients such as lotus seeds, fox nuts and glutinous rice, etc. are stuffed into deboned chicken wings, while retaining the mild, balanced flavours of typical Huaiyang cooking.

材料

全雞翅 _3 隻（500 克）

配料

糯米 _80 克

冬筍 _1 個（150 克）

水發香菇 _2 朵

金華火腿 _30 克

小青豆 _20 克

江瑤柱 _10 克（浸軟）

海參 _20 克（浸發）

紅蘿蔔 _30 克

調味料

生抽 _15 克

蠔油 _20 克

鹽 _5 克

紹興酒 _15 克

糖 _5 克

麻油 _10 克

薑末 _15 克

葱末 _20 克

油 _1.5 公斤

胡椒粉 _2 克

做法

1_ 糯米用水浸泡 3 小時或以上；冬筍、香菇、金華火腿、江瑤柱、海參及紅蘿蔔切成 0.5 厘米小粒。

2_ 全雞翅去骨，洗淨，擦乾水分。

3_ 糯米、各款配料小粒及調味料醃 30 分鐘。

4_ 青豆飛水，與醃好的糯米料混合，釀入全雞翅內（不能釀得太滿），用竹籤封口，放入蒸籠蒸 40 分鐘，取出。

5_ 燒熱油至 180℃，放入雞翅炸至表面金黃酥脆，盛起，瀝乾油分，上碟點綴即可。

Ingredients

3 whole chicken wings (500 g)

Condiments

80 g glutinous rice
1 bamboo shoot (150 g)
2 dried shiitake mushrooms (soaked in water till soft)
30 g Jinhua ham
20 g green peas
10 g dried scallops (soaked in water till soft)
20 g sea cucumber (rehydrated)
30 g carrot

Seasoning

15 g light soy sauce
20 g oyster sauce
5 g salt
15 g Shaoxing wine
5 g sugar
10 g sesame oil
15 g grated ginger
20 g finely chopped spring onion
1.5 kg cooking oil
2 g ground white pepper

Method

1_ Soak glutinous rice in water for at least 3 hours. Drain and set aside. Dice bamboo shoot, shiitake mushrooms, Jinhua ham, dried scallops, sea cucumber and carrot into small cubes about 0.5 cm on one side.

2_ Debone the chicken wingettes while keeping the wing tips and drumettes intact. Rinse and wipe dry.

3_ In a bowl, put in glutinous rice and diced ingredients from step 1. Add seasoning and mix well. Leave them for 30 minutes.

4_ Blanch the green peas. Drain and add to the glutinous rice mixture from step 3. Stuff the wingettes with the mixture, but do not overstuff them. Secure the seam with bamboo skewers or toothpicks. Steam in a bamboo steamer for 40 minutes. Set aside.

5_ Heat oil in a wok up to 180°C. Deep-fry stuffed chicken wings until golden and crispy. Drain. Save on a serving plate. Garnish and serve.

蟹粉獅子頭

"Lion Head" Meatballs with Crabmeat and Crab Roe

獅子頭是淮揚菜最負盛名的菜品之一，
被譽為「國宴第一菜」。
因形似雄獅之頭而得名，
具有肥而不膩、味美湯鮮的特點。

Probably the most famous dish in Huaiyang cuisine, these meatballs are nicknamed "lion heads" because they look like the heads of the male cats with mane. It's also considered the top choice in state banquets, boasting richness, but without being greasy. Its robust meaty flavour is complemented by the flavourful soup.

材料

五花肉 _500 克

配料

蟹粉 _80 克

馬蹄 _150 克

大白菜葉 _4 張

薑片 _15 克

薑末 _5 克

葱段 _25 克

清雞湯 _1 公斤

調味料

紹興酒 _50 克

鹽 _5 克

糖 _2 克

麵粉 _50 克

雞蛋 _1 個

做法

1_ 五花肉切成米粒狀；馬蹄拍碎，剁成小粒，與肉粒、紹興酒 20 克、鹽 4 克、糖 2 克、薑末 5 克、麵粉及雞蛋攪打均勻。

2_ 將肉末擠成六個大丸子，放上蟹粉，放入冰箱冷藏定型。

3_ 砂鍋內放入清雞湯、薑片、葱段及餘下紹酒煮開，加入大肉丸，蓋上大白菜葉，用大火煮開，調至小火燉煮 3 小時，最後灑入少許鹽調味即可。

Ingredients
500 g pork belly

Condiments
80 g crabmeat and crab roe
150 g water chestnuts
4 napa cabbage leaves
15 g sliced ginger
5 g grated ginger
25 g spring onion (cut into short lengths)
1 kg chicken stock

Seasoning
50 g Shaoxing wine
5 g salt
2 g sugar
50 g flour
1 egg

Method
1_ Finely dice the pork belly until it resembles grains of rice. Set aside. Crush the water chestnuts and finely chop them. In a mixing bowl, put in diced pork belly, water chestnuts, 20 g Shaoxing wine, 4 g salt. 2 g sugar, 5 g grated ginger, flour and egg. Mix well.

2_ Squeeze the pork mixture with your hand into 6 large meatballs. Put some crabmeat and crab roe on each meatball. Refrigerate to set their shapes.

3_ In a claypot, pour in chicken stock, sliced ginger, spring onion and the remaining Shaoxing wine. Bring to the boil. Put in the meatballs. Cover with napa cabbage leaves. Turn to high heat and bring to the boil. Turn to low heat and simmer for 3 hours. Sprinkle with salt. Serve.

芋艿清湯鴨

Duck Soup with Yam

每逢中秋佳節，
江南一帶有以芋艿煮鴨的習俗，
以粉糯芋艿與酥嫩鴨肉共煨，
鴨肉酥爛，芋艿軟糯，湯汁清澈，
口味清鮮，色澤明亮。

It's customary for Jiangnan people
to serve this soup on Mid-Autumn Festival.
The yam is starchy and creamy, while the duck is tender and juicy.
The brightly coloured broth is crystal clear with deep umami.

材料

草鴨 _1 隻（約 2 公斤）
毛芋頭 _8 個
水發花菇 _2 朵
熟金華火腿片 _15 克
枸杞 _10 克

配料

薑 _6 片
葱 _3 棵
料酒 _15 克
鹽 _4 克
胡椒粒 _2 克

做法

1_ 草鴨洗淨，放入砂鍋內，加入薑片、葱、胡椒粒及水，蓋過鴨身。

2_ 煮滾後，加入花菇、金華火腿片及料酒，用大火煮 10 分鐘，調至小火燉 3 小時。

3_ 毛芋頭去皮，洗淨，切滾刀塊，放入水煮 10 分鐘，盛起。

4_ 毛芋頭放入鴨湯煮開，用中小火燉 10 分鐘，下鹽調味，最後撒入枸杞即成。

Ingredients

1 duck (about 2 kg)

8 Kamoon yams

2 dried shiitake mushrooms (soaked in water till soft)

15 g cooked Jinhua ham (sliced)

10 g goji berries

Condiments

6 sliced ginger

3 sprigs spring onion

15 g cooking wine

4 g salt

2 g white peppercorns

Method

1_ Rinse the duck and put it into a claypot. Add sliced ginger, spring onion, and peppercorns. Pour in enough water to cover the duck.

2_ Bring to the boil. Add shiitake, Jinhua ham and cooking wine. Boil over high heat for 10 minutes. Turn to low heat and simmer for 3 hours.

3_ Peel the yams. Rinse and cut into random wedges while rolling them on a chopping board. Blanch them in boiling water for 10 minutes. Drain.

4_ Put the yams into the soup. Bring to the boil and turn to medium-low heat. Simmer for 10 minutes. Sprinkle with salt and goji berries. Serve.

農家蒸虎皮肉

Tiger-striped Pork Belly

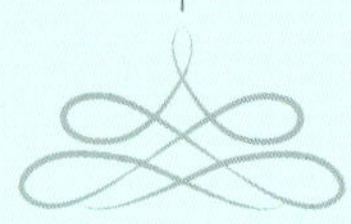

虎皮肉是一道盛載着家的溫馨的經典菜餚，
是逢年過節家家戶戶必備的菜品。
五花肉經過焗煮、煎炸及紅燒等多道工序，
蒸至酥爛而成。

This recipe is a festive favourite that
every family would make around Chinese New Year.
No wonder it's always evokes
the feeling of warmth and comfort.
Pork belly has to be blanched, fried, braised and steamed
to achieve jiggly and melty texture.

材料

帶皮五花肉 _1 公斤

配料

豆苗 _100 克

薑片 _25 克

葱 _20 克

調味料

紹興酒 _60 克

生抽 _15 克

糖 _15 克

做法

1_ 帶皮五花肉用刀刮去皮上細毛，放入熱水飛水，盛起。

2_ 鍋內放入水及五花肉（水蓋過五花肉），加入薑片 15 克、葱 10 克、紹興酒 40 克，用大火煮滾，轉小火慢煮約 30 分鐘，用筷子輕鬆戳入五花肉即可，五花肉瀝乾水分，肉湯保留。

3_ 鍋內下大量油，燒至 200℃，放入五花肉（皮朝下），迅速加蓋，待爆皮聲減小，見五花肉表面呈金黃色，取出，放入煮肉湯內浸泡 15 分鐘，盛起出放涼，切成長 12 厘米、寬 1 厘米長片。

4_ 取直徑 12 厘米的碗，放入餘下薑片及葱，五花肉片皮朝下排入碗內，倒入紹酒、生抽、糖及肉湯 30 克，蒸約 30 分鐘。

5_ 蒸好的五花肉取出，碗口朝下扣入碟內，去掉薑葱，旁邊以炒香的豆苗點綴即可。

Ingredients

1 kg skin-on pork belly

Condiments

100 g pea tips
25 g sliced ginger
20 g spring onion

Seasoning

60 g Shaoxing wine
15 g light soy sauce
15 g sugar

Method

1_ Scrape off the fine hair on the skin of the pork belly with a knife. Blanch in boiling water. Drain and set aside.

2_ Put the pork belly into a pot. Add water to cover. Then put in 15 g of sliced ginger, 10 g of spring onion and 40 g of Shaoxing wine. Bring to the boil over high heat. Turn to low heat and simmer for 30 minutes. Check for doneness by poking the pork with a chopstick. It's done when you can poke through the meat easily. Remove the pork and save the blanching stock for later use.

3_ Pour oil into a wok and heat it up to 200°C. Put the pork belly into the oil with the skin side down. Quickly cover the lid. Wait till the splattering noise subsides. Check the pork to see if it has turned golden. Remove from oil and soak it in the blanching stock from step 2 for 15 minutes. Remove from the stock and leave it to cool. Cut into long strips about 12 cm by 1 cm.

4_ Put the remaining ginger and spring onion into a 12-cm steaming bowl. Arrange the sliced pork belly into the bowl with the skin side down. Pour in Shaoxing wine and light soy sauce. Sprinkle with sugar and 30 g of the blanching stock. Steam for 30 minutes.

5_ Remove the bowl from the steamer and cover it with a serving plate. Turn the bowl upside down to turn out the pork belly. Remove the ginger and spring onion. Stir-fry pea tips in some oil and place them around the pork belly. Serve.

通城藕餅

Tongcheng Fried Lotus Root Stuffed with Pork

通城藕餅作為南通的特色小吃，
是南通文化的重要代表之一。
蓮藕入口脆爽，煎炸後的肉糜鮮香四溢，
藕餅香滑軟糯，老少皆宜，
具有獨特的口感體驗。

It is a characteristic snack in Nantong and
an important part of Nantong's culinary culture.
The lotus root is crunchy while
the pork filling is bursting with meaty flavour.
It is a favourite for all ages that imparts a unique mouthfeel.

材料

蓮藕 _400 克

豬夾心（豬前腿上部）_250 克

配料

薑末 _10 克

葱末 _10 克

麵粉 _100 克

雞蛋 _1 個

油 _500 克

調味料

鹽 _5 克

糖 _3 克

生抽 _3 克

料酒 _20 克

胡椒粉 _ 少許

做法

1_ 蓮藕刨去皮，切成 1.5 毫米夾片。

2_ 豬肉絞成肉蓉，放入薑末、葱末、麵粉 10 克及調味料拌勻。

3_ 蓮藕夾片釀入肉末；麵粉、雞蛋及水調成薄糊狀。

4_ 燒熱油至 100℃，釀蓮藕蘸上薄糊，放入油鍋半煎至兩面金黃，上碟即可。

Ingredients
400 g lotus root
250 g pork picnic shoulder

Condiments
10 g grated ginger
10 g finely chopped spring onion
100 g flour
1 egg
500 g cooking oil

Seasoning
5 g salt
3 g sugar
3 g light soy sauce
20 g cooking wine
ground white pepper

Method
1_ Peel the lotus root. Cut into butterflied slices about 1.5 mm thick.

2_ Grind the pork in a food processor. Add grated ginger, spring onion, 10 g of flour and the seasoning. Mix well.

3_ Stuff each butterflied slice of lotus root with some pork filling. Mix the remaining flour, egg and water into a thin batter.

4_ Heat oil in a wok up to 100°C. Dip the stuffed lotus root into the thin batter one by one. Put into the wok and cook in semi-deep frying manner until both sides golden. Arrange on a serving plate. Serve.

霸王別姬

Hegemon-King Bids His Lady Farewell
(Chicken and Softshell Turtle Soup)

名字取自楚漢相爭「項羽別虞姬」的歷史典故，
以甲魚和雞為主料，傳統做法強調「湯濃肉爛」，
入口即化，口感醇厚，營養豐富。

The name is based on the historical story of Xiang Yu bidding farewell to his wife Consort Yu on the verge of a total defeat and being killed by his enemy. Traditionally, the soup should be condensed while the chicken and turtle should be mushy. This is a nutritious soup with rich textures.

材料

光雞 _1 隻（750 克）
活甲魚 _1 隻（600 克）

配料

冬筍 _50 克
水發乾香菇 _50 克
金華火腿 _40 克
菜心 _3 棵
葱 _20 克
薑 _20 克

調味料

鹽 _10 克
料酒 _50 克
糖 _5 克
清雞湯 _1.5 公斤

做法

1_ 光雞洗淨，飛水，洗淨。

2_ 甲魚宰殺燙洗，刮去黑皮膜，去內臟，飛水洗淨。

3_ 將雞及甲魚放入砂鍋，加入清雞湯，薑、葱、料酒及糖，蒸 1 小時，完成後去掉薑葱，放入冬筍、香菇、金華火腿及菜心，灑入鹽及糖調味，再蒸 5 分鐘即可。

Ingredients

1 dressed chicken (750 g)

1 live softshell turtle (600 g)

Condiments

50 g bamboo shoot

50 g dried shiitake mushrooms (rehydrated)

40 g Jinhua ham

3 sprigs choy sum

20 g spring onion

20 g ginger

Seasoning

10 g salt

50 g cooking wine

5 g sugar

1.5 kg chicken stock

Method

1_ Rinse the chicken. Blanch in boiling water. Rinse again.

2_ Slaughter and dress the softshell turtle. Rinse in boiling water. Scrape off the dark membrane over its skin. Remove all innards. Blanch in boiling water. Rinse well.

3_ Put chicken and softshell turtle into a casserole. Add chicken stock, ginger, spring onion, cooking wine and sugar. Steam for 1 hour. Remove the ginger and spring onion. Add bamboo shoot, shiitake mushrooms, Jinhua ham and choy sum. Sprinkle with salt and sugar. Steam for 5 minutes. Serve.

CHAPTER 5

Dim Sum & Soybean Products

點心及豆品

Dim Sum, bite-size gems that are meticulously crafted, is the treasure trove of Huaiyang cuisine. They are painstaking to make and they boast unique mouthfeels. Crab roe buns are famous for their generous filling that bursts with flavour and the paper-thin wrapper that is barely there. Xiao long bao are known for their beautifully pleated skin and soupy filling. Chicken wontons and shepherd's purse wontons are prized for their fine meaty texture and uniquely earthy and vegetal flavour respectively. The wonton wrapper is velvety, but still toothsome, while the broth is light, but flavoursome.

On the other hand, in the hands of skilful chefs, soybean products like tofu can also be a means to showcase exquisite knife work. Chrysanthemum tofu is sliced finely to resemble the double petaled cluster flower, pleasing not only the eyes, but also the palate. Braised shredded dried tofu is famous for its wispy, hair-like fineness which enhances the way the tofu picks up the rich sauce. And the list goes on. The myriad of dim sum and soupy dishes in Huaiyang cuisine entail unparalleled craftsmanship, emphasising the colour, smell, flavour and form at the same time. They also embody the profound techniques and irresistible charm of the cuisine.

淮揚點心，是淮揚菜的一大瑰寶。以其精細的製作和獨特的口感著稱。其中，蟹黃包以其鮮美的蟹黃餡料和薄如蟬翼的面皮著名；小籠包則以其皮薄餡嫩、湯汁豐富而聞名；雞肉小餛飩與薺菜大餛飩，分別以細膩的雞肉和鮮美的薺菜為餡，餛飩皮滑爽有勁，湯清味美。

菊花豆腐造型宛如綻放的菊花，刀工精細，入口滑嫩；大煮乾絲更以刀工著稱，乾絲細如髮絲，湯汁醇厚，鮮香撲鼻。這些點心與湯菜不僅品種繁多，各具特色，而且製作精細，講究色香味形俱佳，充分展現淮揚飲食的精湛技藝和獨特魅力。

蟹黃鮮肉蒸餃

Steamed Dumplings with Pork and Crab Roe

蒸餃的製作技藝在古代的淮揚地區已流傳，
將蟹黃與鮮肉包入餡製成蒸餃，
其特色在於皮薄如紙，
鮮香獨特。

Ever since the ancient times,
Huaiyang region has been known for
its steamed crab roe and meat dumplings
with paper-thin skin,
juicy savoury filling, and velvety texture.

皮料

中筋麵粉 _100 克
開水 _55 克
鹽 _0.2 克
豬油 _3 克

肉餡

梅花瘦肉 _300 克
肥肉 _200 克
薑 _10 克（與水 120 克打成薑汁）
鹽 _3 克
糖 _10 克
醬油 _25 克
蠔油 _10 克
老抽 _1 克
料酒 _3 克

配料

香菇末 _50 克
紅蘿蔔末 _20 克
杏鮑菇丁 _60 克
麻油 _10 克
蔥油 _10 克
藤椒油 _5 克
花椒油 _2.5 克
胡椒粉 _1 克

蟹黃料

已拆蟹肉 _100 克
菜籽油 _20 克
蔥白、薑末 _ 各少許

做法

1_ 中筋麵粉用開水拌和，加入鹽及豬油，搓成光滑的麵糰，發酵待 20 分鐘。

2_ 蟹黃料拌好；所有餡料放入攪拌盤，拌勻，再加入配料混和。將 250 克肉餡拌入 100 克蟹黃料。

3_ 麵糰用擀麵棍壓平，捲成長條狀，切成每個 8 克小麵糰壓平，包入餡料 15 克，放入蒸爐蒸 7 分鐘即可。

Ingredients

Dumpling skin

100 g plain flour
55 g boiled water
0.2 g salt
3 g lard

Filling

300 g pork shoulder butt
200 g fatty pork
10 g ginger (blend with 120 g water)
3 g salt
10 g sugar
25 g soy sauce
10 g oyster sauce
1 g dark soy sauce
3 g cooking wine

Condiments

50 g diced dried mushroom
20 g diced carrot
60 g diced oyster mushroom
10 g sesame oil
10 g scallion oil
5 g green Sichuan peppercorn oil
2.5 g Sichuan peppercorn
1 g ground white pepper

Crab roe

100 g crab meat
20 g vegetable oil
white part of spring onion
diced ginger

Method

1_ To make the dumpling skin, put plain flour and boiled water into a mixing bowl. Mix well. Add salt and lard. Knead with your hand until the mixture turns into smooth dough. Cover and leave it to rest for 20 minutes.

2_ Mix the crab roe ingredients well. To make the filling, put all ingredients into a mixing bowl. Mix well. Add the condiment and crab roe together.

3_ Roll the dough out with a rolling pin. Then roll it up into a log. Cut into small pieces of dough, each weighing 8 g. Roll out each small piece of dough. Put 15 g of filling over it. Fold and crimp the edges. Steam in a steamer for 7 minutes. Serve.

黃橋燒餅

Huangqiao Scallion Flatbread

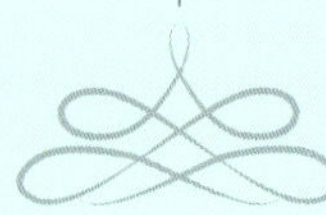

黃橋燒餅是江淮一帶特色小吃之一，
得名於 1940 年著名的戰役——「黃橋決戰」。
曾被選入開國大典的國宴，
先後榮獲「天下第一餅」及「中華第一餅」等稱號。

This is a characteristic snack in Jianghuai region named after the famous Batter of Huangqiao. As a course at the proclamation banquet of People's Republic of China, it is also known as "the world's number one flatbread" and "China's number one flatbread".

水皮料

中筋麵粉 _100 克
豬油 _20 克
水 _55 克
糖 _5 克

油酥料

低筋麵粉 _100 克
豬油 _50 克

餡料

生板油 _100 克
鹽 _2 克
葱 _20 克

塗面料

蛋清 _ 適量
炒香白芝麻 _ 適量

做法

1_ 水皮料及油酥料分別拌和成麵糰，放在室溫發酵 20 分鐘。

2_ 油酥麵糰壓平，放上水油麵糰，摺成三摺，包好，壓平後，再重複一次三摺法，然後捲成長條狀。

3_ 餡料拌勻，混合備用。

4_ 麵糰切成每個 20 克小麵糰，包入餡料 5 克，捏好，表面輕塗蛋清，放上白芝麻點綴。

5_ 放入焗爐以上火 210℃、下火 175℃ 烤 20 分鐘即可。

Ingredients

Water dough

100 g plain flour

20 g lard

55 g water

5 g sugar

Oil dough

100 g cake flour

50 g lard

Filling

100 g pork fat (finely diced)

2 g salt

20 g spring onion

Egg white wash and garnish

egg white (whisked)

toasted white sesames

Method

1_ In two mixing bowls, put in the water dough and oil dough ingredients separately. Mix and knead into dough. Cover and leave them to rest for 20 minutes at room temperature.

2_ Roll out the oil dough. Put the water dough on top. Fold in thirds. Roll it out again. Repeat folding in thirds one more time. Then roll into a log.

3_ To make the filling, put all ingredients into a bowl and mix well.

4_ Cut the dough into small pieces, each weighing 20 g. Roll each piece out and put 5 g of filling on it. Fold and pinch the seam to secure. Brush some egg white on top and sprinkle generously with toasted white sesames.

5_ Bake in an oven with the upper heating element at 210°C, lower heating element at 175°C for 20 minutes. Serve.

草頭煎餅

Pan-fried Burclover Dumpling

草頭學名苜蓿，
在北方被譽為「牧草之王」，
盛載着人們對春天的美好記憶，
有着明確的地域和時令標識。

Burclover is called the king of animal feed in Northern China.
But of course, it's equally nutritious for human consumption.
People often associate burclover with
spring's beauty and its connotation of renewal.
It's a symbol with regional and seasonal importance.

皮料

糯米粉 _250 克
澄麵 _100 克
低筋麵粉 _10 克
草頭葉 _300 克

餡料

草頭末 _100 克
肉丁 _20 克
香菇丁 _10 克
豬油 _30 克
芝麻油 _10 克
鹽 _4 克
糖 _7 克
胡椒粉 _1 克

做法

1_ 皮料的草頭灼水，用水沖後，打成草頭汁，與其他粉料拌和，搓成幼滑麵糰，發酵待 20 分鐘。

2_ 所有餡料放入攪拌盤，拌勻備用。

3_ 發好的麵糰壓平，搓成長條，切成每個 25 克小麵糰。

4_ 小麵糰壓平，包入餡料 10 克，捏成圓餅狀。

5_ 煎鍋下油，放入草頭麵糰，兩面煎 6 分鐘呈金黃即可。

Ingredients

Dumpling skin

250 g glutinous rice flour
100 g wheat starch
10 g cake flour
300 g burclover leaves

Filling

100 g burclover (finely chopped)
20 g diced pork
10 g diced shiitake mushrooms
30 g lard
10 g sesame oil
4 g salt
7 g sugar
1 g ground white pepper

Method

1_ To make the dumpling skin, blanch burclover leaves in boiling water. Rinse and puree. Add all dry ingredients. Mix well and knead into smooth dough. Cover with damp towel and leave it to rest for 20 minutes.

2_ To make the filling, put all ingredients into a mixing bowl. Mix well.

3. Roll the dough out. Then roll into a long log. Cut into small pieces, each weighing 25 g.

4_ Roll out each small piece of dough. Wrap 10 g of filling in it. Gather the edges and secure the seam. Roll into a ball and press gently to flatten slightly.

5_ Heat oil in a frying pan. Put in the dumplings. Fry each side for 3 minutes until golden. Serve.

翡翠燒賣

Jadeite Shaomai

翡翠燒賣是揚州經典著名小吃之一，
其狀如石榴，色如翡翠，
與千層油糕並稱為揚州麵點的「點心雙絕」，
更被評選為江蘇榜「中國地域十大名小吃」。

A classic small eat from Yangzhou,
it's shaped like a pomegranate, and coloured like jadeite.
Together with mille feuille oil cake,
they are known as "the ultimate dim sum duo"
in Yangzhou traditions.
Jadeite shaomai also made it to the list of
top 10 regional famous snacks in China.

皮料

中筋麵粉 _100 克

菠菜葉 _50 克（打成汁）

鹽 _0.5 克

餡料

上海青（小棠菜）_50 克

冬筍丁 _5 克

香菇丁 _5 克

紅蘿蔔末 _2 克

鹽 _1 克

糖 _2 克

油 _1 克

豬油 _1 克

麻油 _0.5 克

做法

皮料

1_ 中筋麵粉加入菠菜汁、鹽拌勻成麵糰，發酵 30 分鐘。

2_ 發好的麵糰壓平，搓成長條，切成每個 7 克小麵糰，做成花邊皮子，備用。

餡料

1_ 上海青放入熱水燙灼，浸於冷水待涼，切碎，壓乾水分。

2_ 放入冬筍丁、香菇丁、紅蘿蔔末拌好，再下調味料拌勻。

3_ 用皮子包好餡料，放入蒸籠以大火蒸 5 分鐘即可。

Ingredients
Shaomai skin

100 g plain flour
50 g spinach leaves (pureed)
0.5 g salt

Filling

50 g Shanghainese bok choy
5 g diced bamboo shoot
5 g diced shiitake mushrooms
2 g finely diced carrot
1 g salt
2 g sugar
1 g cooking oil
1 g lard
0.5 g sesame oil

Method
Shaomai skin

1_ In a mixing bowl, add spinach puree and salt to plain flour. Mix and knead into dough. Leave it to rest for 30 minutes.

2_ Roll out the dough and roll it into a long log. Cut into small pieces, each weighing 7 g. Press with your thumb into thin round disc with wavy edges.

Filling and assembly

1_ Blanch Shanghainese bok choy in boiling water until done. Soak in cold water to cool. Finely chop it and squeeze dry.

2_ Mix Shanghainese bok choy with bamboo shoot, shiitake mushrooms, carrot. Add seasoning and mix again.

3_ Wrap the filling in the shaomai skin. Fold the skin upward and gather on top. Shape each shaomai like a pomegranate. Steam in a steamer over high heat for 5 minutes. Serve.

薺菜餛飩

Shepherd's Purse Wonton

在江浙地區，薺菜有「聚財」諧音之美好寓意，
這是極具特色的傳統美食。
外觀形似元寶，小巧精緻。
餛飩皮軟滑且富有韌性，吃起來清香脆嫩。

In Jiangzhe dialect,
the term for shepherd's purse rhymes with "to gather wealth".
No wonder this wonton is shaped like
a gold ingot as a wish for fortune.
This traditional staple food is small in size and refined in form.
The wonton skin is silky, but resilient.
It tastes soft, but still toothsome.
The unique vegetal flavour of shepherd's purse
also imparts nice aromas.

餛飩皮料
中筋麵粉 500 克
粟粉 30 克
鹽 5 克
蛋清 40 克
水 190-200 克

薺菜豬肉餡料
梅頭豬 250 克
肥肉 250 克
薺菜 250 克（擠乾水分，切碎）
香菇丁 25 克
紅蘿蔔丁 25 克
葱白 30 克

肉餡調味料
鹽 6 克
糖 2.5 克
醬油 10 克
薑 8 克（攪打成汁，與水 120 克拌勻）
胡椒粉 1 克
麻油 3 克
粟米油 15 克
雞蛋 25 克

湯底
水 500 克
雞湯 250 克
鹽 5 克

做法

餛飩皮

1_ 皮料放入攪拌碗，拌成光滑麵糰，發酵待 20 分鐘。

2_ 麵糰擀成薄片，切成 9 厘米方塊（約 12 塊，每塊約 10 克）。

餡料

1_ 梅頭豬肉及肥肉剁碎，拌入調味料混和。

2_ 燒熱油，放入香菇丁及紅蘿蔔丁炒香，加入葱白炒勻，關火，待涼，加入肉末及擠乾水分的薺菜，拌勻後盛起。

3_ 每塊餛飩皮包入 15 克餡料，捏好備用。

製作餛飩

1_ 湯底的水及雞湯煮滾，下鹽調味備用。

2_ 煮滾水 750 克，放入餛飩煮開，加入適量冷水再煮，隨後再次加入適量冷水，待 6 分鐘後，餛飩盛起放入碗內，配上熱雞湯，灑上葱花即可。

Ingredients

Wonton skin

500 g plain flour

30 g corn starch

5 g salt

40 g egg white

190-200 g water

Filling

250 g pork shoulder butt

250 g fatty pork

250 g shepherd's purse (rinsed, squeezed dry, finely chopped)

25 g diced shiitake mushrooms

25 g diced carrot

30 g white part of spring onion (diced)

Seasoning for filling

6 g salt

2.5 g sugar

10 g light soy sauce

8 g ginger (pureed in a food processor, mixed with 120 g of water)

1 g ground white pepper

3 g sesame oil

15 g corn oil

25 g whisked egg

Broth

500 g water

250 g chicken stock

5 g salt

Method

Wonton skin

1_ Put all ingredients into a mixing bowl. Stir and knead into smooth dough. Cover with a damp towel and leave it to rest for 20 minutes.

2_ Roll the dough out into a thin sheet. Cut into 12 squares about 9 cm by 9 cm each, weighing about 10 g each.

Filling

1_ Finely chop the pork shoulder butt and fatty pork. Add seasoning and stir well.

2_ Heat oil in a wok. Stir-fry diced shiitake mushrooms and carrot until fragrant. Add white part of spring onion. Toss well and turn off the heat. Leave it to cool completely. Add pork from step 1 and shepherd's purse that has been squeezed dry. Mix again.

3_ Put 15 g of filling on each piece of square wonton skin. Fold it up and pinch the seam to secure.

Assembly

1_ To make the broth, pour water and chicken stock into a pot. Bring to the boil. Season with salt.

2_ In another pot, boil 750 g of water. Put in the wontons and bring to the boil again. Pour in a bowl of cold water. Bring to the boil again. Repeat adding water and boiling it until the total cooking time has reached 6 minutes. Transfer the wontons into a serving bowl with a strainer ladle. Pour the hot chicken stock from step 1 over the wontons. Sprinkle with finely chopped spring onion. Serve.

水果酒釀元宵

Mini Glutinous Rice Balls with Fresh Fruits in Jiuniang Sweet Soup

酒釀元宵是江南地區的傳統特色小吃，
其特色在於將傳統酒釀與多采的水果巧妙結合，
既保留酒釀的醇厚與甘甜，又融入水果的清新與多汁。

This traditional dessert from Jiangnan region is prized for the drool-worthy combination of a myriad of fresh fruits with the traditional jiuniang sweet soup.
The tangy and juicy fruits accentuate the richness and the mild sour sweetness of jiuniang, without overpowering it.

材料

酒釀 250 克
小元宵（小丸子）280 克
馬蹄粉 30 克
鳳梨 100 克
草莓丁 100 克
雪梨丁 100 克
桃子丁 100 克
冰糖 100 克
桂花醬 40 克
水 1 公斤（其中 200 克勾獻用）

做法

1_ 鳳梨切塊，放入冰糖及水煮 30 分鐘，取出鳳梨塊，糖水留用。

2. 小元宵放入熱水煮 3 分鐘，盛起，過冷水，備用。

3. 鳳梨糖水煮開後，加入馬蹄粉勾獻，放入小元宵煮開，關火，拌入桂花醬及酒釀，裝盤上碗，最後放入水果丁即可。

Ingredients

250 g jiuniang (sweet fermented rice)
280 g mini glutinous rice balls
30 g water chestnut starch
100 g pineapple
100 g strawberries (diced)
100 g ya-li pear (diced)
100 g peach (diced)
100 g rock sugar
40 g candied osmanthus
1 kg water (200 g of which is used to make water chestnut starch slurry)

Method

1_ Cut pineapple into chunks. Cook pineapple in a pot with rock sugar and water for 30 minutes. Set aside the pineapple. Save the syrup for later use.

2_ In another pot, boil water and cook mini glutinous rice balls for 3 minutes. Drain and rinse well.

3_ Boil the pineapple syrup from step 1. Stir in water chestnut starch slurry and cook until it thickens. Put in the blanched mini glutinous rice balls. Bring to the boil. Turn off the heat. Stir in candied osmanthus and jiuniang. Save in serving bowls. Top with diced fruits and pineapple. Serve.

菊花豆腐

Chrysanthemum Tofu

淮揚刀工絕唱的至雅清饌，
以嫩若凝脂的豆腐為紙，浸入琥珀色清雞湯，
瓣瓣纖毫於澄澈中舒展，似白菊湊波吐蕊。

The ultimate manifestation of the exquisite knife work in Huaiyang cuisine,
this recipe uses the delicate and pristine tofu as a blank canvas,
upon which the amber-coloured chicken stock fills the lines and
gaps between the meticulously sliced petals.
The ethereal wisps wave and jiggle in the stock,
like a blooming chrysanthemum with its stamens sticking out to catch pollen.

材料

內脂豆腐 _2 盒

清雞湯 _500 克

配料

鹽 _4 克

糖 _2 克

枸杞 _3 顆

做法

1_ 內脂豆腐打密切成十字花刀，成菊花型。

2_ 清雞湯灑入鹽及糖調味，盛入碗內，放入菊花豆腐蒸 20 分鐘。

3_ 取出蒸好的菊花豆腐，以枸杞點綴即成。

Ingredients

2 boxes gluconolactone (GDL) tofu

500 g chicken stock

Seasoning and garnish

4 g salt

2 g sugar

3 goji berries

Method

1_ Make fine and dense square-shaped cuts on each piece of tofu without cutting all the way through.

2_ Pour chicken stock into a steaming bowl. Add salt and sugar. Put in the chrysanthemum tofu from step 1. Steam for 20 minutes.

3_ Garnish the chrysanthemum tofu with goji berries. Serve.

大煮乾絲

Braised Shredded Dried Tofu

這道菜是揚州傳統名菜，
是淮揚菜中典型的刀工菜品，
被評為江蘇十大經典名菜之一，
口感軟彈、香滑。

This famous dish from Yangzhou is an exemplar of Huaiyang dishes that entail exquisite knife work. It is named one of the top 10 classic recipes of Jiangsu, boasting soft, bouncy and velvety texture.

材料

白乾（豆腐乾）_2 塊
雞胸肉 _25 克
竹節蝦 _5 隻
金華火腿 _15 克
上海小菜心 _5 棵

配料

薑 _2 片
葱 _1 棵
鹽 _18 克
清雞湯 _1.3 公斤（留部分油）
油 _15 克

雞湯料

母雞 _1 隻
瘦肉 _250 克
金華火腿 _15 克
薑 _4 片
葱 _3 棵
水 _400 克
紹興酒 _10 克

做法

1_ 雞洗淨，瀝乾水分。鍋內水 4 公升用大火煮滾，放入薑片、紹興酒、雞、瘦肉和金華火腿，水蓋過面，加蓋以大火煮沸，關火，待 25 分鐘，取出雞待涼。

2_ 取出 25 克雞胸肉放入雪櫃，雞胸肉撕成細長幼絲，其他雞件繼續煮湯，用小火煮 2-2.5 小時，備用。

3_ 金華火腿浸泡清水 1 小時，去掉部分鹽分，沖淨，蒸 10 分鐘，待涼，切絲備用。

4_ 蝦挑出蝦腸，放入煮沸的鹽水內煮開，轉中小火煮 2 分鐘，盛起，泡在冰水備用。

5_ 白乾洗淨，刀沾水，白乾切成薄片，再切成幼絲，泡在鹽水待 30 分鐘，取出沖水，再放入沸水，盛起，沖冷水，重複以上步驟，去除豆腥味，待涼。

6_ 雞湯 650 克及鹽 6 克煮滾，放入白乾絲待 15 分鐘，過濾白乾絲；再煮沸雞湯 650 克及鹽 4 克，放入雞絲煮開，盛起。

7_ 熱鍋放入冷油，放入薑片及葱段爆香，倒入泡白乾絲的雞湯煮開，撇去浮沫，去掉薑葱。

8_ 青菜、蝦仁、雞絲及火腿絲依次放在篩子，放入雞湯內加熱及灼熟。

9_ 白乾絲盛於湯碗，依次放入火腿絲、雞絲及蝦仁，以青菜裝飾，最後澆上清雞湯即可。

Ingredients

2 pieces dried tofu
25 g chicken breast
5 Japanese tiger prawns
15 g Jinhua ham
5 sprigs Shanghainese baby bok choy

Condiments

2 slices ginger
1 sprig spring onion
18 g salt
1.3 kg chicken stock (do not skim off the fat)
15 g cooking oil

Chicken stock

1 hen
250 g lean pork
15 g Jinhua ham
4 slices ginger
3 sprigs spring onion
400 g water
10 g Shaoxing wine

Method

1_ Rinse the chicken. Drain well. Boil 4 litres of water in a pot. Add ginger, Shaoxing wine, chicken, lean pork and Jinhua ham. Make sure the water is enough to cover all ingredients. Cover the lid and bring to the boil over high heat. Turn off the heat and leave it for 25 minutes. Remove the chicken from the stock and let cool.

2_ Carve out 25 g of chicken breast. Refrigerate till cool. Tear the chicken breast into thin shreds. Put the remaining of the chicken back in the stock. Bring to the boil and turn to low heat. Simmer for 2 to 2.5 hours.

3_ Soak Jinhua ham in water for 1 hour to make it less salty. Rinse well and steam for 10 minutes. Leave it to cool. Finely shred it.

4_ Devein the prawns. Blanch them in boiling salted water. Bring to the boil and turn to medium-low heat. Cook for 2 minutes. Drain and soak them in ice water.

5_ Rinse the dried tofu. Wet a knife and slice thinly. Then cut into fine strips. Soak them in salted water for 30 minutes. Rinse. Then blanch them in boiling water. Drain and rinse in cold water. Repeat this blanching and rinsing step to remove any grassy taste of soybean. Drain and leave them to cool.

6_ Boil 650 g of chicken stock from step 2 in a pot. Add 6 g of salt. Turn off the heat and put in the shredded dried tofu. Let it soak in the stock for 15 minutes. Strain and set aside. In another pot, boil 650 g of chicken stock and add 4 g of salt. Put in the shredded chicken breast from step 2. Bring to the boil and drain.

7_ Heat up a wok and add oil. Stir-fry ginger and spring onion until fragrant. Pour in the chicken stock that was used to soak the shredded dried tofu. Bring to the boil and skim off any foam on the surface. Remove the ginger and spring onion.

8_ Put Shanghainese baby bok choy, prawns, chicken breast, and Jinhua ham separately into strainer ladles. Warm them up by inserting the ladles into the hot chicken stock.

9_ Arrange the shredded tofu in a soup bowl. Arrange Jinhua ham, chicken breast and prawns neatly around it. Garnish with Shanghainese baby bok choy. Drizzle with hot chicken stock. Serve.

通派淮揚菜

著者
陳萍

項目統籌
方曉嵐

技術顧問
張錦祥

責任編輯
簡詠怡

翻譯
Wendell J. Leers

攝影
梁細權

裝幀設計
羅美齡

排版
羅美齡、楊詠雯

出版者
萬里機構出版有限公司
香港北角英皇道 499 號北角工業大廈 20 樓
電話：2564 7511　　傳真：2565 5539
電郵：info@wanlibk.com
網址：http://www.wanlibk.com
　　　http://www.facebook.com/wanlibk

發行者
香港聯合書刊物流有限公司
香港荃灣德士古道 220-248 號荃灣工業中心 16 樓
電話：2150 2100　　傳真：2407 3062
電郵：info@suplogistics.com.hk
網址：http://www.suplogistics.com.hk

承印者
中華商務彩色印刷有限公司
香港新界大埔汀麗路 36 號

出版日期
二〇二五年七月第一次印刷

規格
大 16 開（210 mm × 275 mm）

Published in Hong Kong, China.
Printed in China.
ISBN 978-962-14-7591-6

瓷器贊助：
希信（亞洲）有限公司